AF744175

Sugarcane - Biotechnology for Biofuels

Edited by Muhammad Sarwar Khan

Published in London, United Kingdom

IntechOpen

Supporting open minds since 2005

Sugarcane - Biotechnology for Biofuels
http://dx.doi.org/10.5772/intechopen.91550
Edited by Muhammad Sarwar Khan

Contributors
Pietro Sica, Julia M. de O. Camargo, Jhuliana Marcela Gallego Ríos, Graziella C. Antonio, Juliana T.C. Leite, Katia Scortecci, Nathalia Medeiros, Rini Setiati, Aqlyna Fatahanissa, Shabrina Sri Riswati, Septoratno Siregar, Deana Wahyuningrum, safdar mirza, Muhammad Sarwar Khan, Ghulam Mustafa, Faiz Ahmed Joyia

© The Editor(s) and the Author(s) 2021
The rights of the editor(s) and the author(s) have been asserted in accordance with the Copyright, Designs and Patents Act 1988. All rights to the book as a whole are reserved by INTECHOPEN LIMITED. The book as a whole (compilation) cannot be reproduced, distributed or used for commercial or non-commercial purposes without INTECHOPEN LIMITED's written permission. Enquiries concerning the use of the book should be directed to INTECHOPEN LIMITED rights and permissions department (permissions@intechopen.com).
Violations are liable to prosecution under the governing Copyright Law.

cc BY

Individual chapters of this publication are distributed under the terms of the Creative Commons Attribution 3.0 Unported License which permits commercial use, distribution and reproduction of the individual chapters, provided the original author(s) and source publication are appropriately acknowledged. If so indicated, certain images may not be included under the Creative Commons license. In such cases users will need to obtain permission from the license holder to reproduce the material. More details and guidelines concerning content reuse and adaptation can be found at http://www.intechopen.com/copyright-policy.html.

Notice
Statements and opinions expressed in the chapters are these of the individual contributors and not necessarily those of the editors or publisher. No responsibility is accepted for the accuracy of information contained in the published chapters. The publisher assumes no responsibility for any damage or injury to persons or property arising out of the use of any materials, instructions, methods or ideas contained in the book.

First published in London, United Kingdom, 2021 by IntechOpen
IntechOpen is the global imprint of INTECHOPEN LIMITED, registered in England and Wales, registration number: 11086078, 5 Princes Gate Court, London, SW7 2QJ, United Kingdom
Printed in Croatia

British Library Cataloguing-in-Publication Data
A catalogue record for this book is available from the British Library

Additional hard and PDF copies can be obtained from orders@intechopen.com

Sugarcane - Biotechnology for Biofuels
Edited by Muhammad Sarwar Khan
p. cm.
Print ISBN 978-1-83968-935-2
Online ISBN 978-1-83968-936-9
eBook (PDF) ISBN 978-1-83968-937-6

For EU product safety concerns:
IN TECH d.o.o., Prolaz Marije Krucifikse Kozulić 3, 51000 Rijeka, Croatia, info@intechopen.com or visit our website at intechopen.com.

We are IntechOpen, the world's leading publisher of Open Access books

Built by scientists, for scientists

5,300+
Open access books available

131,000+
International authors and editors

155M+
Downloads

156
Countries delivered to

Our authors are among the
Top 1%
most cited scientists

12.2%
Contributors from top 500 universities

WEB OF SCIENCE™

Selection of our books indexed in the Book Citation Index in Web of Science™ Core Collection (BKCI)

Interested in publishing with us?
Contact book.department@intechopen.com

Numbers displayed above are based on latest data collected.
For more information visit www.intechopen.com

Meet the editor

Muhammad Sarwar Khan has a vibrant career in agriculture, education, and biotechnology. He has earned his Ph.D. from the University of Cambridge, UK. The Rockefeller Foundation awarded him a prestigious postdoctoral fellowship under the Rice Biotechnology Program for Developing Countries to research at the Waksman Institute of Microbiology, Rutgers, the State University of New Jersey. His first-of-its-kind research was published in Nature Biotechnology. Dr. Khan served as the founding group leader of Chloroplast Transformation and Biopharming and head of the Biotech Interdisciplinary Division at NIBGE. He is currently serving as the professor of Plant Molecular Biology at the Center of Agricultural Biochemistry and Biotechnology (CABB), University of Agriculture, Faisalabad, Pakistan. Dr. Khan has supervised more than 120 Ph.D. candidates, MPhil students, and researchers who are now serving at national and international levels in various research institutes and universities. He has vastly published in high-impact journals, including Nature, and is the author of several book chapters and books. Dr. Khan has made significant contributions to the field of agricultural biotechnology. He has developed transgenic sugarcane, which was approved by the National Biosafety Committee (NBC) for field trials in 2007. This was the first proposal of indigenously developed GM plants approved by the National Biosafety Committee (NBC) in Pakistan. Dr. Khan has also pioneered plastid transformation in rice and sugarcane, recalcitrant plant species. He has also knocked out several genes from the chloroplast genome of higher plants to assign functions. His current research interests include synthetic biology, whole-genome sequencing, pan-genomics, the development of cost-effective therapeutics, and edible vaccines for animals. Dr. Khan has received prestigious national and international awards, honors, and medals. Currently, he is serving as editor/guest editor on several international research journals and books, published by various publishers such as Springer, Bentham, Taylors & Francis, and IntechOpen. Dr. Khan has attended, as speaker and session chair, several international workshops and conferences including the Nobel Prize Summit 2021, organized by the National Academies of Sciences, Engineering, and Medicine.

Contents

Preface

Sugarcane (Saccharum officinarum) is one of the most vegetatively cultivated cash crops in the world, predominantly grown in tropical and subtropical regions in more than 90 countries, and it is the main source of sugar, biofuel, and biomaterial. Despite enormous efforts, crop production is compromised due to several biotic and abiotic factors; therefore, the crop demands more attention from breeders and biotechnologists to improve its sustainable production. In this era of depleting natural energy resources, biotechnology has played a major role in converting sugarcane into energy-cane (ethanol- and lipid-cane) to meet the requirements of energy on a sustainable basis.

The book Sugarcane - Biotechnology for Biofuels consists of seven chapters that are strategically organized, allowing readers to follow the contents easily. It begins with Chapter 1, "Introductory Chapter: Bioengineered Sugarcane - A Sustainable Biofactory of Renewable Energy", in which Dr. Khan comprehensively highlights how engineered sugarcane provides superior genotypes with regulated expressions of genes and improved enzymatic reactions for increased bioconversion of lignocellulose biomass, and recovery of sugars and lipids for ethanol and diesel production. Further, he demonstrates how cellulose and lignin could be used to develop high-quality fiber and biosurfactants for the apparel and petroleum industries, respectively. In Chapter 2, "Base Excision Repair in Sugarcane – A New Outlook", Drs. Medeiros and Scortecci explain the association of the base excision repair (BER) pathway with the maintenance of genome integrity in sugarcane and other model plants by using different bioinformatic tools. They follow-up with questions on the components and evolutionary aspects of the BER pathway in plants. Dr. Sica, in Chapter 3, "Sugarcane Breeding for Enhanced Fiber and Its Impacts on Industrial Processes", discusses the advancements in the breeding programs to improve biomass and fiber rather than the sucrose contents of cane crops to exploit the potential of advanced genotypes to meet the energy demands. Dr. Leite and colleagues, in Chapter 4 very comprehensively highlight the physicochemical properties of residues from the sugarcane industry such as straw, bagasse, vinasse, and filter cake, aiming at their use in energy recovery processes. In Chapter 5, Dr. Khan and his team discuss the advancements as well as the potential challenges in the production of sugarcane biofuel by focusing on genetic and genomic interventions to improve the crop as energy-cane. In Chapter 5, they also discuss the controversies in the production and usage of biofuel derived from sugarcane. Finally, in Chapter 6, Dr. Setiati and coauthors highlight the potential of bagasse as a rich source of lignin for the development of lignosulfonate surfactant, an anionic surfactant to improve the oil recovery in the petroleum sector.

Muhammad Sarwar Khan, Ph.D.
Center of Agricultural Biochemistry and Biotechnology (CABB),
University of Agriculture,
Faisalabad, Pakistan

Chapter 1

Introductory Chapter: Bioengineered Sugarcane - A Sustainable Biofactory of Renewable Energy

Muhammad Sarwar Khan

1. Introduction

Engineered sugarcane provides superior genotypes with a regulated expression of endogenous and exogenous genes, improved enzymatic reactions and sugar production, resulting in increased bioconversion of lignocellulosic biomass, recovery of sugars, and lipids for ethanol and diesel production.

The genus *Saccharum* belongs to the family *Poaceae* with five major species; *officinarum*, *sinense*, *barberi*, *robustum,* and *spontanuem*, developed through complex hybridization. The *Saccharum officinarum* is mainly grown for sugar in tropical and subtropical regions of the world. Where this crop meets up to 80% of world sugar requirement [1] is facing several problems including biotic and abiotic that affect its production. Conventional approaches lagged in solving the problems due to the complex genome and narrow genetic pool of sugarcane, however, advanced tools are employed to address biotic and abiotic problems, and to increase agronomic traits including yield, juice, and sugar. Further, improved sugarcane is being grown for bioethanol production in countries like the USA, Brazil, European Union, Guatemala, and China. Recently, sugarcane is engineered not only for bioethanol also for oil, hence it opens a window to develop purpose-grown sugarcane in the world. The chapter specifically highlights the technologies exploited in improving the crop for sustainable production of biofuel, including ethanol and diesel.

Though fossil fuels meet approximately 80 percent of the energy demand of the world and are required for economic development, the system is not renewable and causing global warming due to heavy CO_2 emissions. Global warming may affect the agenda of sustainable development therefore it is imperative to address this issue. The development of modern approaches adds to our confidence in addressing these objectives in the energy system.

2. Sugarcane genetic improvement

Sugarcane has several unique characteristics compared to other cereals. Sugarcane is highly polyploid at its genome level, carries photosynthesis through C_4 mechanism, chloroplast distribution in mesophyll and bundle sheath cells are few unique traits of this monocotyledonous plant. These traits of sugarcane make this plant one of the valuable specimens to explore its genetic potential. Despite considerable varietal improvement using conventional interventions, a serious

IntechOpen

effort to develop elite genotypes is required. Major hurdles in the genetic improvement of sugarcane are due to its genetic complexity, low fertility, long cycle, and climate-specific flowering nature of the crop. Under this scenario, biotechnology can play a leading role to improve and introduce commercially important traits into elite genotypes.

Starting from the somaclonal selection under *in vitro* conditions to genetically engineering the genome of the crop each of the approaches has made it possible to develop new traits in sugarcane; including insect resistance, herbicide- and abiotic stresses-tolerance. Interestingly, with the introduction of synthetic elements, the pathways have been engineered in the sugarcane, resulting in various industrial and non-industrial products. Briefly describing the technology, each cell carries three major organelles; namely, nucleus, chloroplast, and mitochondria. Each of these organelles has its genome where the nuclear genome is routinely manipulated to develop new traits by exploiting two major approaches of gene transfer; *Agrobacterium*-mediated and biolistic-mediated gene gun methods. Of these two methods, particle bombardment is more successful to develop transgenic plants in sugarcane [2]. Several steps are involved to complete the process of genetic modification of the genome, starting from the choice of a gene to develop a trait, selection of a promoter and a terminator to regulate the expression of the trait-conferring gene, selection of a selectable marker for selection and purification of transformed cells, selection of an explant and optimization of regeneration medium by fine-tuning the concentration of the growth regulators, supporting the cell to complete cycle of differentiation and regeneration into a shoot.

To transform the chloroplast genome, the *plastome*, requirements of the transformation process are different. In this process, the antibiotics that normally kill prokaryotic organisms are preferred, keeping in view the mode of actions of the drugs. For example, spectinomycin, an aminoglycoside antibiotic is used to select and purify the transgene-recipient cells on regeneration medium for several plants including members of families; Solanaceae, *Brassicaceae,* Apiaceae, and Asteraceae; including; tobacco, potato, tomato, cabbage, oil rapeseed, carrot, and lettuce. However, monocotyledonous crops of the Poaceae family, like, sugarcane and rice are resistant to spectinomycin, naturally; therefore, another antibiotic namely, streptomycin is added to the regeneration medium for selecting the transplastomic clones. Chloroplast transformation in sugarcane has been reported recently [3] using a vector carrying fluorescent antibiotic resistance marker gene, FALRE-S [4]. The transformed plants remained heteroplasmic after successive cycles of selection and regeneration owing to the complex anatomy and polyploidy of the plastome, though, homoplasmic clones have been successfully obtained in several plant species [5]. The sugarcane plastome is very attractive when talks about metabolic engineering where high-level expression of genes and accumulation of the product is required.

2.1 Genetic improvement for insect resistance

Insects are serious threats to the sugarcane industry, worldwide, though exact assessment of economic losses is difficult to record. However, fragmented information is available on individual pests. Major pests of the crop are chewing, sucking insects, and canegrubs, and termites. Amongst borers, the most common are top, stem, and root borers.

Engineering the genome for enhanced resistance to insects is one of the success stories of transgenic technology. Several molecules including proteinase inhibitors (PI), secondary metabolites, ribosome-inactivating proteins, lectins, and δ-endotoxins of *Bacillus thuringiensis* [6–8], have been identified for effectively controlling the insects. Therefore, developing insect-resistant sugarcane through a transgenic approach is required. Several transgenic sugarcane plats have been developed in different regions of the world keeping in view the type of insects invading the crop and resulting in economic losses. Successful examples are the development of transgenic plants addressing the problems of top borers using δ-endotoxins of *Bacillus thuringiensis* in Indian subcontinent, American and African countries. The larvae invade the sugarcane crop two times, at early and at a later stage, during the growing season. Early infestation causes serious damage to the crop by developing a 'dead heart', young shoot died as larvae chew the base of the shoot while later benefits to increase population and feed, triggering yield loss from 15 to 30 percent. How the process of 'dead heart' starts, the young larvae tunnels into the nucleus of the spindle and damage the growing point, causing the shoot to wilt and die. This requires the development of resistance against several factors including biotic and abiotic in sugarcane [9], but traditional improvement strategies are hampered by high polyploidy, genetic complexity, environment-specificity, and low fertility [10]. Thus biotechnological approaches have the potential to address such stresses by engineering resistance. Since the first genetic manipulation of sugarcane [11, 12], the development of transgenic clones is a routine [13, 14]. Several traits including effective control of stem borers have been introduced into the sugarcane [15, 16]. In these studies, a synthetic *cry1Ab* gene was selected to introduce and express under a tissue-specific promoter, PEPC. After bombardment the transformed cells, subsequently shoots were recovered on phosphinothricin-containing regeneration medium. Expectedly, the primary clones accumulated varied amounts of expressed CRY protein, ranging from 20 to 40 ng/mg. The variation in expressed protein levels is perhaps due to the random insertion of the transgene, varied copy number of the transgene in transgenic clones. Once, homozygous clones were recovered, the varied expression was documented as developmental and photosynthetic control of expression of the transgene. However, the expression levels were regarded as the highest levels of toxin reported so far in the literature. These levels were 13- and 35-fold higher than the highest levels of modified *cry1Ac* and *cry1Ab* gene, respectively [16–18]. The maize ubiquitin promoter was used to drive the synthetically developed cry1Ac gene in sugarcane for stem borers [16, 17], which is five to six times higher compared to the levels obtained by a constitutive promoter, CaMV 35S, in sugarcane getting high expression under PEP-C promoter were perhaps of using a C4 plant-specific promoter.

Interestingly, the detectable endotoxins levels varied from base to tip and from first to the outermost leaf of the whorl, depending on the developmental stage of the leaf. Measurements of the toxin levels in leaves showed that the accumulation of transprotein was low in young emerging leaves and increased with the leaf development. The maximum amount was recorded in the fully developed leaf on the same plant [18]. In another investigation, the *cry1Ac* gene has been introduced using a vector-dependent method for borer-resistance in sugarcane. The transgene was successfully introduced and transformed plants showed a high level of toxicity to *Sesamia cretica* giving 100% mortality of the larvae. These two reports are success stories of developing borer-resistant transgenic sugarcane [19]. Other success stories of developing transgenic sugarcane with different genes used for effective control of diverse insects are not included intentionally due to the word limit of the article, hence, my apologies for not including such valuable published reports.

2.2 Genetic improvement for disease resistance

Sugarcane is vulnerable to several diseases caused by different organisms including; bacteria, fungi, viruses, and nematodes. The diseases damage the crop throughout the growing season and multiple epidemics by 64 diseases are reported that have damaged the sugarcane crop, worldwide. Of these diseases, five are caused by bacterial and 40 by fungal pathogens. However, the most common are rust, wilt, red rot, and smut that have seriously affected the sugarcane, demanding the development of disease-resistant varieties. In this section, only two diseases; one of fungal and the other of bacterial origin, are discussed.

Amongst fungal diseases, red rot is the most common disease of sugarcane and is caused by *Colletotrichum falcatum* Went that attacks the sucrose accumulating parenchyma cells [20] of culmus stalk of sugarcane, causing severe losses; up to 29% in the cane yield and 25 to 75% in the sugar recovery [21, 22]. Since sugarcane is a vegetatively propagated plant hence the pathogens spread through diseased-culmus-setts, stressing the availability of healthy canes for propagation. Instead of chemical, biological control of the red rot in sugarcane using *Trichoderma* (*Trichoderma harzianum and Trichoderma viride*) is reported [23, 24]. Of these two species, *Trichoderma harzianum* is more effective in controlling the *Colletotrichum falcatum*. It is interesting to know how *Trichoderma* which is a fungus controls another fungus *Colletotrichum*? Such invading fungi control others through the mechanisms of mycoparasitism and antibiosis. Mycoparasitism is one of the most important mechanisms for biocontrol of pathogen fungi that work through three steps: chemotrophic growth and recognition; coiling and interaction of hyphae and secretion of specific lytic enzymes [25]. Whereas, antibiosis is a shift to the concept of mycoparasitism that is based on lethal principle. Trichoderma produces low molecular weight and diffusible substances that penetrate the host cell thereby inhibit the uptake of nutrients, sporulation, production of metabolites, and the synthesis of the cell wall of the target fungus [26, 27]. Nevertheless, antibiosis is a species-specific mechanism.

Transgenically, the red rot was controlled by expressing genes from *Trichoderma* into the genome of sugarcane. Encouraged from the inhibitory effects of β1,3-glucanase, isolated from *Trichoderma*, against pathogenic fungi, Nayyar *et al.*, [28] expressed the gene in sugarcane and analyzed the regenerating plants for transgene integration and expression. The transgene was expressed up to 4.4-fold higher than the non-transformed wild-type plants, and resistance to two pathotypes of *Colletotrichum falcatum*. The expression levels of the transgene were up-regulated after infection compared to the levels recorded before infection in engineered resistant sugarcane plants. However, pathogenicity tests on transgenic sugarcane against the virulent strains of *Colletotrichum falcatum* will demonstrate the resistance level. To demonstrate the pathogenicity, the transgenic sugarcane plants were challenged with two virulent strains (CF08 and Cf09) of *Colletotrichum falcatum*, and plants exhibited moderate to high-level resistance against Cf09 and Cf08, respectively. The structural model of the β-1,3-glucanase encoded protein demonstrated that two active sites namely; Glutamate 628 and Aspartate 569 of the catalytic domain of the protein are the main sites that have catalyzed the cleavage of β-1,3-glycosidic bonds and lysis of pathogen hyphae [29].

Further, another disease caused by a bacterial pathogen in sugarcane, the leaf scald, is also a serious threat to the sugarcane industry as it severely affects the yield. The disease is mainly distributed in the Philippines, Thailand, Myanmar, Vietnam, Java, Laos, Australia, and USA [30, 31] but now it is the most important quarantine disease in Taiwan, Guangxi, Fujian, Jiangxi, Yunnan, Guangdong, and Hainan in China [32]. The leaf scald, also called leaf burning disease which is caused by *Xanthomonas albilineans* infects the xylem vessels of the sugarcane plant. Recently, it is reported that it also invades tissues of leaves and culmus stalks, predominantly sucrose accumulating cells. Unlike other phytopathogenic bacteria, *Xanthomonas albilineans* lacks a type 3 HRP secretion system, a system that enables most bacteria to overcome the defense mechanism of their host plants [32]. Yet, it is capable of invading other tissues, and developing symptoms in three stages namely; latent, chronic, and acute phases. It is observed that the bacterium remains dormant for more than 12 months without any symptoms. The bacterium infection develops symptoms of chlorotic stripes and patches on leaves, side shooting starts at the base of culmus stalks and burning of the leaf tips, and well-defined white pencil-lines along the veins during the chronic phase. Sometimes, the sudden death of whole stools.

Biotechnological interventions to address diseases are primarily focused on engineering the sugarcane genome and molecular breeding. The complete genome has been sequenced despite the complexities of ploidy levels, individual genes are transferred to engineer resistance traits. The coding region of an albicidin detoxifying gene (*albD*) from *Pantoea dispersa* has been expressed under *ubi* promoter from the maize and the *nos* terminator using Agrobacterium into the nuclear genome of the sugarcane plants found susceptible to leaf scald [33]. However, the transgenic plants were resistant to the disease since no chlorotic disease symptoms appeared in inoculated leaves, whereas non-transgenic plants developed severe symptoms. Further, a high concentration of accumulated transprotein has protected young stems against the multiplication of the pathogen. Thus, transgenic sugarcane clones conferred resistance to the disease as the plants showed no symptoms and multiplication of the bacterium [34, 35]. Introducing transgenes into sugarcane represents an important step forward, however, the growing investment in genomics and identification of resistance gene analogs (RGA), present in a majority of indigenous genotypes, might become an attractive alternative in disease resistance management through transgenic technology, and the development of screening tools of resistant genotypes.

2.3 Genetic improvement for sugar production and recovery

Sugarcane leaves contain both the mesophyll and bundle sheath cells. Photosynthesis occurs in the chloroplasts of both cells. Sugarcane unlike other organisms accumulates photosynthetic assimilates in the form of sucrose in the culmus stalk, thereby directly providing sugar for human consumption. Sucrose produced in the leaves is translocated from the source (leaves) to sink (parenchyma cells of the culmus stalk) through the phloem. The sucrose-loving parenchyma cells of the stalk accumulate the sucrose to an exceptionally high concentration i.e. up to 25–27% of the fresh weight [36, 37]. However, there is a correlation between photosynthetic activity and the sucrose contents in sugarcane. Toward maturity of the cane, the sucrose contents in the stem are increased whereas photosynthesis in leaves is reduced [38], indicating the regulatory mechanism of sucrose accumulation in the stalk cells [39]. Sucrose synthesis, transport, and accumulation in sugarcane are continuous processes.

Sucrose is accumulated in the stalk for metabolism and storage, and this accumulation is believably dependent on source supply, the storage capacity of the sink, and the metabolism in the parenchyma cells of the cane stem. Conceptually, the sucrose supply and demand situation determines whether the plant is source-limited or sink-limited. Accordingly, when source supply is limited in a plant to meeting the demand of the sink, then it will be a source-limited plant and when the demand is less than the source supply then it will be the sink-limited plant [40]. Generally, the photosynthetic activity falls with the maturation of the culmus stalk, and the rate of photosynthesis in high sucrose-accumulating elite lines is reduced to two-third of low sucrose accumulating lines, suggesting the role of the source-sink communication in the sucrose accumulation [38, 41]. For example,

Saccharum spontaneum accumulates sucrose at a lower level with high photosynthesis compared to the noble canes [42].

Sugarcane has developed a special mechanism of carbon fixation, and fix CO_2 in the form of four carbon molecules. This carbon fixation is facilitated by the coordination of cell types; mesophyll and bundle sheath cells, and hence photosynthetic functions are divided between these two cells. Carbon dioxide is initially fixed in the mesophyll cells, where it is converted into bicarbonate by phosphoenolpyruvate carboxylase. The product formed is oxaloacetate, which is reduced to malate by the NADP-dependent malate dehydrogenase. The malate is transported to the bundle sheath cells, located near the vascular system of the leaf, where it is converted to pyruvate, releasing CO2 in the vicinity of the ribulose-1,5-bisphosphate carboxylase-oxygenase (RuBisCO). In bundle sheath cells the carbon dioxide is fixed by RuBisCO to energy-rich molecules such as glucose. After this step, the Calvin cycle proceeds as in C3 plants. Knowledge of C4 photosynthesis has made it possible to understand sugarcane productivity and highlights its importance to become one of the principal sources of carbohydrates for human consumption and bioethanol production.

Sucrose synthesized in the leaves of sugarcane has to be accumulated into the parenchyma cells of a stalk, experimentally it has been documented that sucrose is transported mainly through the symplastic system [43–45]. However, it is imperative to understand the distribution and accumulation of sucrose in three compartments namely; apoplast, cytoplasm, and vacuole to comprehensively overview the accumulation of sucrose in the parenchyma cells [46]. The transported sugar is unloaded into the apoplast and then transported to vacuoles. This transport of sugar to vacuoles is carried out by the sucrose transporters under low turgor conditions. Another way to translocate the unloaded sucrose is that the sucrose is hydrolyzed into glucose and fructose in the apoplast by the cell wall acid invertase and the end products (hexoses) are transported by hexose carriers, allowing the continued efflux of sucrose from the phloem. In this context, the cooperation of both the hexose transporters and the cell wall invertase is important and has been extensively reviewed elsewhere [47]. In plants sucrose is hydrolyzed by two enzymes; the sucrose synthase and cell wall invertase. Of these two enzymes, the sucrose synthase reversibly converts sucrose into glucose and fructose for utilization in respiration and cellulose biosynthesis [48], invertase irreversibly converts sucrose into glucose and fructose, exerting a pivotal role in carbon utilization and distribution.

Sucrose starts accumulating in the internodes at the early growth stage but the accumulation sharply increases when internode elongation stops [49], coinciding with the development of the stem and its elongating internodes [50]. Further, elongating internodes are characterized by a high ratio of hexoses to sucrose [51]. Higher concentrations of hexose in the developing internodes could be due to the predominant apoplasmic phloem unloading pathway as a result of sucrose cleavage by cell wall invertase and cellular uptake by hexose transporters [52]. Experimentally it has been demonstrated in radiolabeled sucrose transport studies that the majority of the translocated sucrose is taken into parenchyma cells of both developing and mature internodes, extensively reviewed [53], despite cell wall invertase activity was present in the apoplasm. Initially, vacuoles in the parenchyma cell of sugarcane stalk function as a sink to store sucrose through growth and development-dependent reversible process, but at a later stage when elongation of internodes stops then mature culm stores sucrose in the apoplast, intercellular spaces outside the plasma membranes.

Through the interventions of genetic engineering approaches, several attempts have been made to improve the sucrose contents of the sugarcane genotypes.

Sucrose synthase and cell wall invertase are the main regulatory enzymes of sucrose metabolism. When sucrose synthase was overexpressed in sugarcane, the contents were not improved [54]. Similarly, cell wall invertase was regulated to improve the sucrose contents of the sugarcane but no significant increase in sucrose accumulation was observed [55]. Though sucrose contents were increased in sugarcane cell suspension cultures by suppressing the invertase [56] yet it was not reproducible in stable transgenic sugarcane plants [57], indicating the role of the regulatory feedback mechanism between source and sink during accumulation [58]. Recently, two homologs of invertase inhibitors are identified in sugarcane, believed to be members of the superfamily of pectin methylesterase inhibitor, moderately conserved in plants [59]. To demonstrate where both enzymes are precisely located in the plant, both genes were fused with the green fluorescent protein and experimentally observed that ShINH1 is targeted to the apoplast. Further, expressed tissue specifically and is developmentally regulated, suggesting its role in metabolic regulation of sucrose between source and sink during sucrose accumulation in sugarcane. The gene expresses at relatively high levels in leaves and stalk but decreases significantly in stalk toward the maturity of the internodes. Experimentally, ShINH1 potently inhibited acid invertase, making it a candidate for controlling the deterioration of sucrose in sugarcane. However, experimental and *in silico* studies have revealed that both ShINH1 and ShINH2 have a role in sucrose accumulation and may contribute to the improvement of sugar yield and recovery in sugarcane.

2.4 Genetic improvement for bioconversion of biomass

Commercially, the juice is extracted from sugarcane and processed through multistep procedures to develop sugar, ethanol, or biodiesel, leaving fibrous material, bagasse, in a large volume of approximately 540 million metric tons on yearly basis, worldwide, which is used in sugar mills as a furnace fuel and apparel industry for paper or board manufacturing, however, its use as a feedstock for cellulosic ethanol is gaining importance and several independent reports have been published on its chemical composition, constituents preparation, and utilization. Sugarcane bagasse consists of 41 to 55 percent cellulose, 20 to 27.5 percent of hemicellulose, 18 to 26.3 percent of lignin, and around 7 percent of others by weight. Of these three major constituents, cellulose and lignin have gained much attention to develop byproducts, and/or for value addition [60].

Bioconversion of lignocellulosic biomass from bagasse has gained much attention of the plant biotechnologists and microbiologists for bioethanol production and value-added products as bagasse has long been considered in bioethanol industries for ethanol production, referred to as 'second-generation ethanol. Biotechnological interventions have enabled the efficient bioconversion of bagasse through the introduction of improved microbial strains, media formulations, and product recovery processes. As the bagasse has a complex structure, therefore it is pretreated to dissociate the lignin-cellulose and to increase the surface area for better enzymatic activities to convert the biomass to fermentable sugars. Different approaches including; biological treatment, dilute acid hydrolysis, alkali hydrolysis, and solvent-based pretreatment have been reported for *saccharification* of the bagasse. However, all such processes significantly add to the cost of bioethanol production from lignocellulosic biomass [61].

The saccharification process is affected by the high contents of lignin and lignin syringyl to guaiacyl (S/G) ratio in the bagasse hence the composition of the lignin is needed to be changed as they prevent cellulase from accessing the cellulose molecules in the process of ethanol production [61]. Biotechnological approaches may help in altering the composition of this fibrous material by regulating the genes

(*Saccharum* species). Frontiers in Plant Science. 2013;**4**:201-206

[48] Coleman HD, Yan J, Mansfield SD. Sucrose synthase affects carbon partitioning to increase cellulose production and altered cell wall ultrastructure. Proceedings of the National Academy of Sciences USA. 2009;**106**:13118-13123

[49] Hawker JS, Hatch MD. Mechanism of sugar storage by mature stem tissue of sugarcane. Physiologia Plantarum. 1965;**18**:444-453

[50] Bowen JE, Hunter JE. Sugar transport in immature intermodal tissue of sugarcane II. Mechanism of sucrose transport. Plant Physiology. 1972;**49**:789-793

[51] Julius BT, Leach KA, Tran TM, Mertz RA, Braun DM. Sugar transporters in Plants: New Insights and Discoveries. Plant Cell Physiology. 2017;**58(9)**:1442-1460

[52] Ruan YL. Sucrose metabolism: gateway to diverse carbon use and sugar signaling. Annual Review of Plant Biology. 2014;**65**:33-67

[53] Tarpley L, Vietor DM. Compartmentation of sucrose during radial transfer in mature sorghum culm. BMC Plant Biology. 2007;7:33-43

[54] Vickers JE, Grof CPL, Bonnett GD, Jackson PA, Morgan TE. Effects of tissue culture, biolistic transformation, and introduction of PPO and SPS gene constructs on performance of sugarcane clones in the field. Crop Pasture Science. 2005;**56**:449-456

[55] Rossouw D, Bosch S, Kossmann J, Botha FC, Groenewald JH. Down-regulation of neutral invertase activity in transgenic sugarcane cell suspension cultures leads to a reduction in respiration and growth and an increase in sucrose accumulation. Functional Plant Biology. 2007;**34**:490-498

[56] Ma H, Albert HH, Paull R, Moore PH. Metabolic engineering of invertase activities in different sub-cellular compartments affects sucrose accumulation in sugarcane cells. Australian Journal of Plant Physiology. 2000;**27**:1021-1030

[57] Botha FC, Sawyer BJB, Birch RG. Sucrose metabolism in the culm of transgenic sugarcane with reduced soluble acid invertase activity. Congress Proceedings-International Society of Sugar Cane Technology. 2001;**24**:588-591

[58] Chandra A, Jain R, Rai RK, Solomon S. Revisiting the source–sink paradigm in sugarcane. Current Science. 2011;**100**:978-980

[59] Shivalingamurthy SG, Anangi R, Kalaipandian S, Glassop D, King GF, Rae AL. Identification and Functional Characterization of Sugarcane Invertase Inhibitor (ShINH1): A Potential Candidate for Reducing Pre- and Post-harvest Loss of Sucrose in Sugarcane. Frontiers in Plant Science. 2018;**9**:598-611

[60] Arapova OV, Chistyakov AV, Tsodikov MV, Moiseev II. Lignin as a Renewable Resource of Hydrocarbon Products and Energy Carriers (A Review). Petroleum Chemistry. 2020;**60**:227-243

[61] Fan LT, Gharpuray MM, Young-Hyun L. Evaluation of Pretreatments for Enzymatic Conversion of Agricultural Residues. Biotechnology and Bioengineering Symposium. 1981;**11**:29-45

[62] Jung JH, Fouad WM, Vermerris W, Gallo M, Altpeter F. RNAi suppression of lignin biosynthesis in sugarcane reduces recalcitrance for biofuel production from lignocellulosic

involved in its biosynthesis pathway. Three key genes; namely, *COMT* (caffeic acid *O*-methyltransferase), *CCoAOMT* (caffeoyl-CoA *O*-methyltransferase), and *F5H* (ferulate 5-hydroxylase) have been targeted employing techniques; RNAi (RNA interference), TALEN (transcription activator-like effector nuclease), and CRISPR (Clustered Regularly Interspaced Short Palindromic Repeats)/case9. The gene, *COMT* was downregulated from the biosynthesis pathway of lignin using the RNAi approach, and the expression of the gene was lowered by 67–97%, and the lignin contents were reduced by 3.9% to 13.7%, respectively. Further, the S/G ratio in the lignin was reduced from 1.47 in the wild type to values ranging between 1.27 and 0.79. Consequently, levels of fermentable sugar were increased up to 29% without pretreatment however, these levels were further increased by 34% when biomass was treated with dilute acid [62]. Further, COMT mutant lines developed by targeting the conserved region of the gene through the TALEN approach showed lignin reduction up to 19.7% with 43.8% improved saccharification efficiency. The lignin contents were further reduced from 29 to 32% in COMT mutants [63]. Owing to the highly complex and polyploid genome, targeted mutagenesis using CRISPR/Cas9 could be a valuable tool to reduce the lignin contents.

As far as the bioconversion of the lignocellulosic biomass to fermentable sugars is concerned, significant quantities of cellulolytic and hemicellulolytic enzymes are required, and currently, microbes are used to produce these enzymes that have substantially increased the production cost of bioethanol. Therefore, *in planta* production of enzymes could have a significant impact on the economics of bagasse-based ethanol production. Several independent reports have been published describing the hydrolysis of pretreated biomass by mixtures of purified cellulases from wheat and barley straw, corn stover, switchgrass, and poplar. Afterward, genes encoding the enzymes like; cellobiohydrolase I (CBH I), cellobiohydrolase II (CBH II), and endoglucanase (EG) were stably introduced into the genome of the sugarcane, and the accumulation of enzymes was confirmed in the leaves of plants [64]. However, the enzymes CBHI and CBHII of fungal origin remained resistant to proteolysis during sugarcane leaf senescence, while bacterial origin (EG) was degraded, demonstrating the stability of recombinant cellulase in transgenic sugarcane. When tissue-specific regulatory elements, with or without targeting sequences, were used to control the expression of transgenes the accumulation of enzymes was variably enhanced. All three enzymes accumulated to higher levels when targeted to vacuoles however, the highest levels of accumulation of endoglucanase were recorded when the enzyme was targeted to the chloroplasts. This practically demonstrates a significant step forward for the economic production of lignocellulosic ethanol. Hence, such initiatives could advance cellulosic ethanol technology.

2.5 Genetic improvement for ethanol production

Renewable fuels are an attractive alternative to petroleum-derived gasoline since it potentially lowers the production of greenhouse gases, and add to the country's economy. Amongst biofuels, bioethanol is being produced from non-food as well as food parts of the plants including sugarcane. It is prepared from can juice, molasses, and bagasse. The renewable bioethanol industry has been successfully developed in countries like the USA, Brazil, Mexico, the EU, and China, and bioethanol is being used to unravel the energy crisis. Developing bioethanol from non-food sources has emerged as a trend due to several advantages like availability of abundant raw materials, low price, and renewability. Such bioethanol industry is strategically being developed in China.

Bioethanol production in the USA is from corn with a maximum production of 56.7 billion liters from cereal grains, though the focus has been shifted recently from corn to cellulosic ethanol, facing many challenges like feedstock availability and high conversion costs. Brazil, the largest sugarcane-based ethanol producer in the world is pioneering in using bioethanol as a motor fuel. In 2019–2020, Brazilian ethanol production reached 34 billion liters to meet the domestic market demand. To increase the target of 50 billion liters of ethanol the Brazilian industry is increasing its fermentation efficiency from 83–90%. This will enable the ethanol industry to meet the environmental targets established by Brazil at COP21, The *United Nations Climate Change Conference*, which was held in Paris, France, from 30 November to 12 December 2015. In the market, it is being sold in the form of either pure ethanol fuel (E100) or blended with gasoline (E27). Presently, ethanol production from sugarcane produces nine times more energy than the energy consumed during its production process [65]. Hence, bioethanol is produced directly from the sugarcane juice converting this into ethanol through the fermentation process using microorganisms as sugarcane juice is rich in sucrose, glucose, and fructose. During the fermentation process, the sucrose is readily broken down into glucose and fructose by the yeast invertase. During the process, the sugarcane stalk is crushed and milled with water. The extracted juice is heated at a temperature of 115°C. After heating, the juice is treated with either sulfuric acid or lime, and the excessive inorganic compounds are precipitated. Afterward, the heated juice is cooled down and yeast is added along with nutrients for yeast growth. Fermentation can be carried out in either batch or continuous reactors, however, continuous reactors are used in Brazil.

Biotechnological attempts have been made to improve the sugarcane for agronomic traits and recently, several interventions have been made to improve the genetic makeup of the plant as well as of microorganisms used to develop ethanol from sugarcane. The *Saccharomyces cerevisiae* is commonly used in the fermentation process of sugarcane to develop ethanol. One of the potential candidate genes associated with ethanol fermentation is PHO4, a phosphate regulon [66]. The replacement of the gene from fast-growing strain MC15 to a high ethanol-producing strain MF01, exploiting the homologous recombination approaches, has improved the ethanol yield to 5.30% as the maximum yield harvested was 114.71 g/L and has decreased the fermentation time to 12.5%, compared to the non-recombinant MF01 strain. So, the engineered strain will not only be improving the yield of the ethanol also reducing the fermentation time. In another experiment, the F-514 strain of the *S. cerevisiae* was used to ferment sugars from sugarcane molasses in hot and dry weather, and it was reported that by improving the fermentation parameters ethanol yield was improved. In addition to yeast strains, the pretreatment of molasses or can juice with lime or sulfuric acid affect the ethanol yield in the fermentation process. Raharja et al. [67] used commercial instant dry yeast to simplify the production process and reduce the bacterial contamination risk. Using the molasses as starting material, containing a very high concentration of sugar (30%), and pretreated with sulfuric acid produced ethanol at a very low level, lower than the levels attained using bagasse as feedstock. Hence, producing ethanol from feedstock like sugars and molasses and its application as an energy source may not be desirable on a long-term basis as the demand for fuel, food, and feed is rising, demanding alternative means of cost-competitive and sustainable supply of feedstock. Further, the bioethanol producing industries in the USA as well as in Brazil have decided to develop second-generation ethanol (2G) from lignocellulose materials, requiring new production models [68]. Biomass, which is burnt to produced energy in sugar industries could be used as feedstock to convert polysaccharides into ethanol [68]. The pretreatment of biomass to extract hemicellulose is carried out either by biological or chemical (diluted acids) treatment or through steam. Bioethanol is produced from hemicellulosic hydrolysate using yeast, immobilized on magnetic particles, in the fermenters under a magnetic field. It has been reported that ethanol production was 34% higher in bioreactors assisted with axial rather than by transversal magnetic field.

2.6 Genetic improvement for oil production

Biofuel's largest markets in the world are in the USA, Brazil, the European Union, and China, collectively producing 85% of global biofuel with a 48% share of the USA alone. Of the total biofuel production, 82% share is of bioethanol, which is majorly produced by the USA and Brazil. However, biodiesel is the second majorly produced biofuel in the world, with a 49% contribution from the European Union. The biodiesel contribution of the EU, USA, and Brazil is 34.1%, 19.5 and 12%, respectively. The feedstock used to extract oil is rapeseed and or cooking oil in the EU whereas in the USA and Brazil is soybean. The soybean has been grown on 33 million hectares in the US with per hectare oil production from 0.36 to 0.61 MT/ha [USDA, 2015]. However, total biodiesel production would have been only about one-tenth of the US distillate fuel oil consumption (about 155 million MT) by crushing the total soybean produced [69]. This situation demands the introduction of other crops, and interventions of recombinant technology to engineer metabolic pathways in plants to accumulate oil, and could be grown on marginal land avoiding any possible competition with major food and feed crops. Metabolic pathways have been engineered in tobacco and Arabidopsis, highly regarded model plants used in genetic engineering experiments, to accumulate triacylglycerol [70]. In another independent study three genes namely; WRI1 (WRINKLED1), DAGA T1–2 (diacylglycerol acyltransferase1–2), and OLE1(Oleosin1) were expressed, and the triacylglycerol was accumulated in leaves and culmus stalk of sugarcane [71]. The accumulation of triacylglycerol added to the total fatty acid contents of up to 4.7% and 1.7% of dry weight in mature leaves and stems, respectively. Interestingly, confocal micrographs have shown the presence of lipid droplets within the transgenic mesophyll cells, indicating a step forward in the accumulation of high levels of triacylglycerol in sugarcane. In another independent study, multiple genes were either expressed or suppressed in sugarcane to increase the lipid contents in the vegetative biomass. The genes CYSOLE1, DAGAT1–2, OLE1, WRI1 were co-expressed, whereas, tgd1 and sdp1 were simultaneously suppressed in the sugarcane and an elevated amount of the TAG was recorded. The transgenic plants with constitutively co-expressed CYSOLE1, DAGAT1–2, OLE1, WRI1, and simultaneously suppressed tgd1 in different plasmids elevated the TAG accumulation by 277-fold and 109-fold in leaf and stem tissues, respectively [71]. Constitutive co-expression of CYSOLE1, DAGAT1–2, WRI1, and co-suppression of tgd1 and sdp1 in a single construct elevated the TAG content by 404-fold in leaves. These findings need further confirmation of TAG accumulation with large-scale field testing. Developing dual-purpose feedstock by either overexpressing or suppressing the genes to produce both ethanol and biodiesel is becoming a routine. Transgenic sugarcane plants were developed by co-expressing genes, namely; WRINKLED1, DGAT1–2, Oleosin1, and suppressing the AGPase and PXA1 genes using the RNAi approach. The AGPase encodes ADP-glucose pyrophosphorylase while PXA1 is a subunit of the peroxisomal ABC transporter1. The engineered plants accumulated triacylglycerol to 5% higher than the non-transformed plants [69]. The TGA levels were 31–33% of the total lipid contents in the transgenic plants. From these canes (**Figure 1**), both sugars and lipids were extracted where the sugar extraction efficiency was as high as 90% with repeated hot water while of lipids was 60%, demanding an improved process to increase the lipid extraction efficiency and for commercialization.

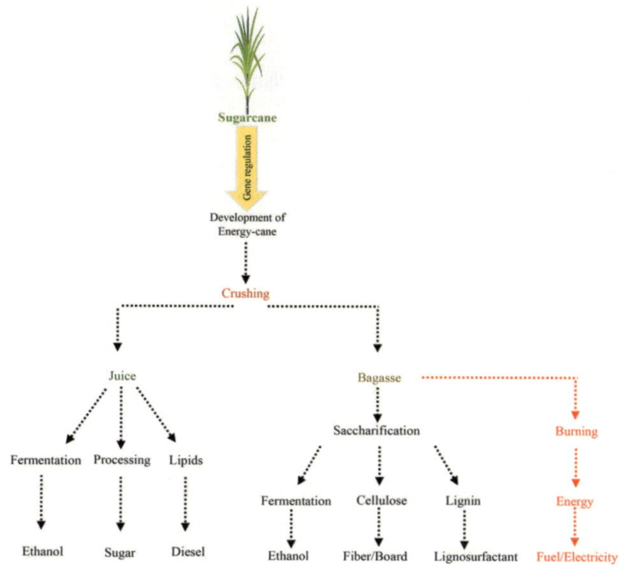

Figure 1.
Schematic representation of improvement of sugarcane genotype for renewable energy and byproducts.

3. Conclusion

The sugarcane genome is highly polyploid with specialized leaf anatomy, the Kranz anatomy where mesophyll cells are arranged around the bundle sheath cells. Mesophyll cells are better connected to the environment and perform functions of phosphoenolpyruvate carboxylase (PEPC). Whereas, bundle sheath cells are rich in chloroplasts to carry out functions of Ribulose-1,5-bisphosphate carboxylase-oxygenase (rubisco) to fix CO_2 to synthesize sugars in the Kelvin cycle. The CO_2 concentration mechanism in sugarcane allows the development of the highest annual yield of biomass. Plant biotechnologists have engineered sugarcane for agronomic as well as value addition traits. Hence, the yield of fermentable sugars could be enhanced by improving the translocation and by bioconversion of lignocellulosic biomass for bioethanol and biodiesel production, sustainably. Further, Cellulose and lignin could be separated and used to develop high-quality fiber and lignosurfactants, respectively.

Author details

Muhammad Sarwar Khan
Center of Agricultural Biotechnology and Biochemistry (CABB), University of Agriculture, Faisalabad, Pakistan

*Address all correspondence to: sarwarkhan_40@hotmail.com

IntechOpen

© 2021 The Author(s). Licensee IntechOpen. This chapter is distributed under the terms of the Creative Commons Attribution License (http://creativecommons.org/licenses/by/3.0), which permits unrestricted use, distribution, and reproduction in any medium, provided the original work is properly cited.

References

[1] OECD/FAO (2020). "OECD-FAO Agricultural Outlook", OECD Agriculture statistics (database), http://dx.doi.org/10.1787/agr-outl-data-en.

[2] Khan MS, Ali S, Iqbal J. Developmental and photosynthetic regulation of *Bacillus thuringiensis* δ-endotoxin reveals that engineered sugarcane conferring resistance to 'dead heart' contains no toxins in cane juice. Molecular Biology Reports. 2011;**38**:2359-2369

[3] Mustafa G, Khan MS. Transmission of engineered plastids in Sugarcane, a C_4 monocotyledonous plant, reveals that sorting of preprogrammed progenitor cells produce heteroplasmy. Plants. 2021;**10**:26-38

[4] Khan MS, Maliga P. Fluorescent antibiotic resistance marker to track plastid transformation in higher plants. Nature Biotechnology. 1999;**17**:910-915

[5] Daniell H, Khan MS, Allison L. Milestones in chloroplast genetic engineering: An environmentally friendly era in biotechnology. Trends in Plant Science. 2002;7:84-91

[6] Weng LX, Deng HH, Xu JL, Li Q, Zhang YQ, Jiang ZD, Li QW, Chen JW, Zhang LH. Transgenic sugarcane plants expressing high levels of modified cry1Ac provide effective control against stem borers in field trials. Transgenic Research. 2011;**20**:759-772

[7] Falco MC, Silva-Filho MC. Expression of soybean proteinase inhibitors in transgenic sugarcane plants: Effects on natural defense against Diatraea saccharalis. Plant Physiology and Biochemistry. 2003;**41**:761-766

[8] Setamou M, Bernal JS, Legaspi JC, Mirkov TE, Legaspi BC. Evaluation of lectin-expressing transgenic sugarcane against stalkborers (Lepidoptera: Pyralidae): Efects on life history parameters. Journal of Economic Entomology. 2002;**95**:469-477

[9] Pandey P, Irulappan V, Bagavathiannan MV, Senthil-Kumar M. Impact of combined abiotic and biotic stresses on plant growth and avenues for crop improvement by exploiting physio-morphological traits. Frontiers in Plant Science. 2017;**8**:537-551

[10] Ithape DM, Maharana M, Tripathy SK. Scope of genetic transformation in sugarcane: a review. Genomics and Applied Biology. 2017;**8(1)**:1-7

[11] Chowdhury MKU, Vasil KI. Stably transformed herbicide resistant callus of sugarcane via microprojectile bombardment of cell suspension cultures and electroporation of protoplasts. Plant Cell Reports. 1992;**11**:494-498

[12] Bower R, Elliot AR, Potier BAM, Birch RG. High efficiency, microprojectile-mediated co-transformation of sugarcane, using visible or selectable markers. Molecular Breeding. 1996;**2**:239-249

[13] Liu D, Oard SV, Oard JH. High transgene expression levels in sugarcane (*Saccharum officinarum* L.) driven by the rice ubiquitin promoter RUBQ2. Plant Science. 2003;**165**:743-750

[14] Weng LX, Deng H, Xu J-L, Li Q, Wang LH, Jiang Z, Zhang HB, Li Q, Zhang LH. Regeneration of sugarcane elite breeding lines and engineering of stem borer resistance. Pest Management Science. 2006;**62**:178-187

[15] Birch RG, Bower R, Elliot A, Potier B, Franks T, Cordeiro G.

Expression of foreign genes in sugarcane. In: Cock JH, Brekelbaum T (eds) Proceedings of the XXII Congress of the International Society of Sugarcane Technologists, Tecnicana, Cali. Vol. 2;1995. p.369-373

[16] Davina Michel D, Bachelier B, Drean JY, Harzallah O. Preparation of Cellulosic Fibers from Sugarcane for Textile Use. In: Conference Papers in Science, Hindawi. Vol. 2013;2013. p.1-6

[17] Weng LX, Deng H, Xu JL, Li Q, Wang LH, Jiang Z, Zhang HB, Li Q, Zhang LH. Regeneration of sugarcane elite breeding lines and engineering of stem borer resistance. Pest Management Science. 2006;**62**:178-187

[18] Jabeen R, Khan MS, Zafar Y, Anjum T. Codon optimization of cry1Ab gene for hyper expression in plant organelles. Molecular Biology Reports. 2010;**37**:1011-1017

[19] Dessoky IS, Roba M, Ismail RM, Elarabi NI, Abdelhadi AA, Abdallah NA. Improvement of sugarcane for borer resistance using Agrobacterium mediated transformation of cry1Ac gene. GM Crops and Food. 2021;**12(1)**:47-56

[20] Uys L, Botha FC, Hofmeyr JHS, Rohwer JM. Kinetic model of sucrose accumulation in maturing sugarcane culm tissue. Phytochemistry. 2007;**68**:2375-2392

[21] Moore PH. Temporal and spatial regulation of sucrose accumulation in the sugarcane stem. Functional Plant Biology. 1995;**22**:661-679

[22] Viswanathan R, Samiyappan R. Red rot disease of sugarcane: major constraint for Indian sugar industry. Sugar Cane. 1999;**5**:9-15

[23] Singh V, Joshi BB, Awasthi SK, Srivastava SN. Ecofriendly management of red rot disease of sugarcane with Trichoderma strains. Sugar Tech. 2008a;**10**:158-161

[24] Singh V, Srivastava SN, Lal RJ, Awasthi SK, Joshi BB. Biological control of red rot disease of sugarcane by Trichoderma harzianum and *T. viride*. Indian Phytopathology. 2008;**61**:486-493

[25] Atanasova L, Crom SL, Gruber S, Coulpier F, Seidl-Seiboth V, Kubicek CP, Druzhinina, IS. Comparative transcriptomics reveals different strategies of Trichoderma mycoparasitism. BMC Genomics. 2013;**14**:121-125

[26] Lorito M, Mach RL, Sposato P, Strauss J, Peterbauer CK, Kubicek CP. Mycoparasitic interaction relieves binding of the Cre1 carbon catabolite repressor protein to promoter sequences of the ech42 (endochitinase-encoding) gene in Trichoderma harzianum. Proceedings of the National Academy of Sciences USA. 1996;**93**:14868-14872

[27] Howell CR. The role of antibiosis in biocontrol. In: Harman GE, Kubicek CP, eds. Trichoderma and Gliocladium. Enzymes, biological control and commercial applications. London: Taylor & Francis. Vol. 2;1998. p.173-184

[28] Nayyar S, Sharma BK, Kaur A, Kalia A, Sanghera GS, Thind KS, Yadav IS, Sandhu JS. Red rot resistant transgenic sugarcane developed through expression of β-1,3-glucanase gene. PLoS ONE. 2017;**12(6)**:e0179723

[29] Vazquez-Garciduenas S, Leal-Morales CA, Herrera-Estrella A. Analysis of the beta-1,3-Glucanolytic System of the Biocontrol Agent Trichoderma harzianum. Applied and Environmental Microbiology. 1998;**64(4)**:1442-1446

[30] Saumtally A, Dookun-Saumtally A, Rao GP, Saumtally AS, Rott P. Leaf scald of sugarcane: A disease of worldwide importance. In: Sugarcane Pathology: Bacterial and Nematode Diseases. New Hampshire: Science Publishers; 2004. p.65-76

[31] Zhang R, Shan H, Li W, Cang X, Wang X, Yin J, Luo ZM, Huang YK. First report of sugarcane leaf scald caused by Xanthomonas albilineans in the province of Guangxi, China. Plant Disease. 2017;**101(8)**:1541.

[32] Huang YK., Li WF., Zhang RY., Wang XY. Diagnosis and Control of Sugarcane Important Diseases. In: Color Illustration of Diagnosis and Control for Modern Sugarcane Diseases, Pests, and Weeds. Springer, Singapore; 2018. p.1-103

[33] Zhang L, Xu J, Birch RG. Engineered detoxification confers resistance against a pathogenic bacterium. Nature Biotechnology. 1999;**17**:1021-1024

[34] Govindaraju M, Li Y, Zhang M. Emerging Bacterial Disease (Leaf Scald) of Sugarcane in China: Pathogenesis, Diagnosis, and Management. In: Plant Diseases-Current Threats and Management Trends. Intech Open; 2019

[35] Birch RG. Xanthomonas albilineans and the anti-pathogenesis approach to disease control. Molecular Plant Pathology. 2001;**2(1)**:1-11

[36] Moore PH. Temporal and spatial regulation of sucrose accumulation in the sugarcane stem. Australian Journal of Plant Physiology. 1995;**22**:661-679

[37] Bull TA, Glasziou KT. The evolutionary significance of sugar accumulation in Saccharum. Australian Journal of Biological Sciences. 1963:**16**:737-742

[38] McCormick AJ, Watt DA, Cramer MD. Supply and demand: sink regulation of sugar accumulation in sugarcane. Journal of Experimental Botany. 2009:**60**:357-364

[39] Watt DA, McCormick AJ, Govender C, Carson DL, Cramer MD, Huckett BI, Botha FC. Increasing the utility of genomics in unravelling sucrose accumulation. Field Crops Research. 2005;**92**:149-158

[40] Patrick JW, Botha FC, Birch RG. Metabolic engineering of sugars and simple sugar derivatives in plants. Plant Biotechnology Journal. 2013;**11**:142-156

[41] Patrick JW. Phloem unloading: sieve element unloading and post-sieve element transport. Annual Review of Plant Physiology and Plant Molecular Biology. 1997;**48(1)**:191-222

[42] Kaffka SR, Grantz DA. Encyclopedia of Agriculture and Food Systems. 2014.

[43] Rae AL, Perroux JM, Grof CP. Sucrose partitioning between vascular bundles and storage parenchyma in the sugarcane stem: a potential role for the ShSUT1 sucrose transporter. Planta. 2005;**220(6)**:817-825

[44] Lalonde S, Tegeder M, Throne-Holst M, Frommer WB, Patrick JW. Phloem loading and unloadin of sugar and amino acids. Plant, Cell and Environment. 2003;**26**:37-56

[45] Grof CPL, Campbell JA. Sugarcane sucrose metabolism: scope for molecular manipulation. Australian Journal of Plant Physiology. 2001;**28**:1-12

[46] Sacher JA, Hatch M, Glasziou KT. Sugar Accumulation Cycle in Sugar Cane. III. Physical & Metabolic Aspects of Cycle in Immature Storage Tissues. Plant Physiology. 1963;**38(3)**:348-354

[47] Wang J, Nayak S, Koch K, Ming R. Carbon partitioning in sugarcane

biomass. Plant Biotechnology Journal. 2012;**10**:1067-1076

[63] Jung JH, Altpeter F. TALEN mediated targeted mutagenesis of the *caffeic acid O-methyltransferase* in highly polyploid sugarcane improves cell wall composition for production of bioethanol. Plant Molecular Biology. 2016;**92:**131-142

[64] Harrison MD, Geijskes RJ, Lloyd R, Miles S, Palupe A, Sainz MB, Dale JL. Recombinant cellulase accumulation in the leaves of mature, vegetatively propagated transgenic sugarcane. Molecular Biotechnology. 2014;**56(9)**:795-802

[65] Altieri A. Bioethanol Development in Brazil. In: Sayigh A, editor. Comprehensive Renewable Energy. Elsevier; 2012. p.15-26

[66] Ogawa N, DeRisi J, Brown PO. New components of a system for phosphate accumulation and polyphosphate metabolism in Saccharomyces cerevisiae revealed by genomic expression analysis. Molecular Biology Cell. 2000;**11(12)**:4309-4321

[67] Raharja R, Murdiyatmo U, Sutrisno A, Wardani AK. Bioethanol production from sugarcane molasses by instant dry yeast. 2019; IOP Conf. Ser.: Earth Environ. Sci. 230 012076

[68] Pereira SC, Maehara L, Machado CMM, Farinas CS. 2G ethanol from the whole sugarcane lignocellulosic biomass. Biotechnology Biofuels. 2015;**8:**44-59

[69] Huang H, Moreau RA, Powell MJ, Wang Z, Kannan B, Altpeter F, Grennan AK, Long SP, Singh V. Evaluation of the quantity and composition of sugars and lipid in the juice and bagasse of lipid producing sugarcane. Biocatalysis and Agricultural Biotechnology. 2017;**10**:148-155

[70] Cai Y, Goodman JM, Pyc M, Mullen RT, Dyer JM, Chapman KD. Arabidopsis SEIPIN Proteins Modulate Triacylglycerol Accumulation and Influence Lipid Droplet Proliferation. Plant Cell. 2015;**27(9)**:2616-2636

[71] Zale J, Jung JH, Kim J, Pathak B, Karan R, Liu H, Chen X, Wu H, Candreva J, Zhai Z, Shanklin J, Altpeter F. Metabolic engineering of sugarcane to accumulate energy-dense triacylglycerols in vegetative biomass. Plant Biotechnology Journal. 2016;**14(2)**:661-669

Chapter 2

Base Excision Repair in Sugarcane – A New Outlook

Nathalia Maíra Cabral de Medeiros
and Katia Castanho Scortecci

Abstract

The base excision repair (BER) pathway has been associated with genome integrity maintenance. Owing to its central role, BER is present in all three domains of life. The studies in plants, considering BER, have been conducted using *Arabidopsis* and rice models. Therefore, future studies regarding BER are required in other organisms, particularly in crops such as sugarcane, to understand its mechanism, which may reflect the uniqueness of DNA repair in monocots. Our previous results have revealed that sugarcane is an interesting plant for studying this pathway considering the polyploidy genome and genome evolution. This chapter aimed to characterize the BER pathway in sugarcane by using different bioinformatics tools, for example, screening for BER homologs in the sugarcane genome to identify its members. Each sequence obtained was subjected to structural analysis, and certain differences were identified when *Arabidopsis* was compared to other monocots, including sugarcane. Moreover, ROS1, DEM, and DML3 were not identified as a complete sequence in the sugarcane EST database. Furthermore, FEN1 is present as two sequences, namely FEN1A and FEN1B, both featuring different amino acid sequence and motif presence. Furthermore, FEN1 sequence was selected for further characterization considering its evolutionary history, as sequence duplication was observed only in *the Poaceae* family. Considering the importance of this protein for BER pathway, this sequence was evaluated using protein models (3D), and a possible conservation was observed during protein–protein interaction. Thus, these results help us understand the roles of certain BER components in sugarcane, and may reveal the aspects and functions of this pathway beyond those already established in the literature.

Keywords: BER, *Saccharum* spp., DNA repair, *Poaceae*, 3D-model, phylogenetic

1. Introduction

The base excision repair (BER) pathway is linked to the maintenance of genome integrity since BER is an essential genome defense pathway, which acts over a broad range of DNA lesions induced by endogenous or exogenous genotoxic agents [1]. Owing to its central role, BER is present in all three domains of life [2]. As a complex process, BER initiated by the excision of damaged base, proceeds through a sequence of reactions that generate various DNA intermediates and finish with the repair of the initial DNA structure. Nevertheless, BER focuses on repair, deals with DNA demethylation and erases the epigenetic mark 5-methycytosine (5mC) and converts it to cytosine [3]. Thus, an emerging and crucial role of BER in epigenetic regulation is being investigated and characterized [4–7]. Although various studies have been conducted in animal and microbial systems, BER knowledge regarding plants has been neglected.

Despite these apparent differences in plant research compared to other organisms, knowledge about the BER pathway in plants has gained immense interest in recent years. The results obtained so far reveal that plants possess orthologues of most BER genes previously found in other organisms [8–10]; however, they also retain some plant-specific BER proteins as well as distinct enzyme combinations not observed in other kingdoms (review by [8]). Unfortunately, most of these findings were based on the model *Arabidopsis thaliana*, indicating the importance of amplifying studies on other organisms, particularly important crops [11].

Grasses (*Poaceae*; alternative name *Gramineae*) are undoubtedly an important plant group considering the economic perspective, and provide essential cereals such as *Eragrostis*, *Hordeum*, *Oryza*, *Secale*, *Sorghum*, *Triticum* and *Zea*; stalks such as *Arundo* and *Phragmites*; cane for food and materials for construction such as *Bambusa* and *Phyllostachys* and sugar crops such as *Saccharum* and *Sorghum* [12]. Sugarcane is a crop of noticeable value that can meet the requirements of food, feed fiber, and fuel. Moreover, sugarcane production by weight surpasses that of food crops such as wheat, rice and maize [13]. Despite its importance, this crop has been given less attention in scientific research than other members of *Poaceae* family, such as rice and maize. One reason is the polyploid and heterozygous nature of its genome, leading to lesser research compared to the other grass species studied [14–16].

Furthermore, research has been conducted using the sugarcane expressed sequence tags (ESTs) project (SUCEST), which has identified possible DNA repair genes [17, 18]. BER sequences were predicted, although these investigations were conducted more than 10 years ago [19]. Since then, there have been several improvements in bioinformatics tools as well as in sugarcane genome sequencing [20–24].

More studies are required to unravel the specific features of BER pathway in sugarcane, which may reflect the uniqueness of DNA repair in monocots. A new screening for BER homologs in the sugarcane genome was developed to gain advanced knowledge of BER in this crop. Each sequence was structurally analyzed. Thereafter, some of these sequences were selected for further investigating their evolutionary history. Tri-dimensional models have also been created to verify the conservation of mechanisms and protein–protein interactions in sugarcane BER components. The intriguing results displayed in this chapter raise questions regarding the roles of certain components of BER in sugarcane, just as in monocots, and they might broaden the aspects and functions of this pathway beyond those already established in the scientific literature.

2. Identification of base excision repair's components in sugarcane

The BER components were identified in sugarcane through homology with the bioinformatic tools. In this regard, the SUCEST-FUN database, which assembles distinct sugarcane databases such as the Sugarcane Expressed Sequence Tags genome project (SUCEST-FUN) (http://sucest.lad.ic.unicamp.br/en/) [25]; Sugarcane Gene Index (SGI); SUCAST catalogs and SUCAMET, which include expression data (http://sucest-fun.org); GRASSIUS database [26] and records of the agronomic, physiological and biochemical characteristics of sugarcane cultivars, were used.

DOI: http://dx.doi.org/10.5772/intechopen.95878

BER component	% Identity (Sb/At)	Protein name	Substrate or function	Role in BER	SUCEST-FUN ID
DEM, ROS1 and DMLE	(94.54/54.88, 51.79, 55.05)*	DEMETER, Repressor of silencing 1 and DEMETER-like protein 3	5-methylcytosine (5-meC)	Involved in the initial stage of BER, recognizing the damaged base.	comp89337_c0_seq1
OGG1	(96.64/59.29)	8-oxoguanine-DNA glycosylase 1	8-oxoguanine (8-oxoG)		comp78469_c0_seq3
NTH1	(97.85/ 54.09, 49.31)**	Endonuclease III homolog 1	oxidized pyrimidine		comp79344_c0_seq1
UDG	(96/57)	Uracil-DNA glycosylase	Uracil		comp64547_c0_seq4 and SCEQFL5048B07.g
MBD4L	(88.3/47)	Methyl-CpG-binding domain protein 4-like protein	G:T mismatches within methylated and unmethylated CpG sites. Uracil or 5-fluorouracil in G:U mismatches.		comp78687_c0_seq2
MUTM (1 and 2)****	(94.49/68.64) (86.83/59.11)	Formamidopyrimidine-DNA glycosylase	oxidation products of 8-oxoguanine (8-oxoG)		comp85541_c0_seq1; SCCCLR2C01B12.g
ARP1 (1 and 3)****	(95.09/59.60) (97.42/71.79)	DNA-(apurinic or apyrimidinic site) endonuclease	Ap site	Repair by-products (AP site) of BER or oxidation.	comp79331_c0_seq10; comp86134_c0_seq5
FEN1 (A and B)****	(96/81.69) (82.97/73.07)	Flap endonuclease 1	5′ flap	Involved with the BER's long-patch.	comp79282_c1_seq1 and SCEPRZ1008D03.g; comp85461_c0_seq2 and comp79282_c1_seq1
Pol λ	(95.16/54.70)	DNA POLIMERASE LAMBDA	Resynthesize missing nucleotides	It replaces the Polymerase beta acting on the BER short-patch.	comp80417_c0_seq9
TDP1	(95/45.7)	Tyrosyl-DNA phosphodiesterase 1	Processing of diverse 3′- and 5′-blocking groups at DNA ends	Processing of intermediate BER products	comp89039_c1_seq9 and comp89039_c1_seq3
LIG1	(91.67/70.96)	DNA ligase 1	Seal 5′-PO4 and 3′-OH polynucleotide ends	Involved in the long and BER's short patch.	comp86584_c0_seq5 and comp86584_c0_seq7
LIG4	(97.53/73.25)	DNA ligase 4		Proposed to be involved in BER's short-patch.	comp85403_c0_seq6

BER component	% Identity (Sb/At)	Protein name	Substrate or function	Role in BER	SUCEST-FUN ID
ZDP	(94.8/43.2)	Polynucleotide 3′-phosphatase ZDP	3′-phosphopolynucleotide	Processing of intermediate BER products.	comp78030_c0_seq1
PCNA	(100/85.55 and 86.69) ***	PROLIFERATING CELL NUCLEAR ANTIGEN	A scaffold to recruit the proteins involved in DNA replication, DNA repair, chromatin remodeling, and epigenetics	Involved in the BER's long-patch.	comp82119_c0_seq2 and SCCCCL3140F04.g
PARP1	(96.7/60.6)	Poly [ADP-ribose] polymerase 1	Uses NAD+ as a substrate, synthesizes and transfers ADP-ribose onto aspartic and glutamic acid residues of acceptor proteins	Protects the BER substrate, present in the BER's long-patch.	comp82301_c0_seq8 and SCAGLB1070H02.g
PARP2	(95.5/53.1)	Poly [ADP-ribose] polymerase 2		Not essential for DNA repair in the BER pathway.	comp85410_c0_seq3 and SCJFRT1012D11.g
XRCC1	(97.11/48.2)	X-RAY REPAIR CROSSCOMPLEMENTING PROTEIN 1	interacting with APE1 and stimulating its AP endonuclease activity, prepares the DNA substrate for the DNA polymerase activities.	Involved in the BER's short-patch.	comp81667_c0_seq2, SCVPFL4C09E05.g and SCEZSD1082B05.g
WRN	(93.9/41.5)	WERNER SYNDROME ATP DEPENDENT HELICASE	Helicase enzyme	Interacts with several BER proteins: FEN1, PolB and PARP1	comp74108_c0_seq1, SCUTAM2089F01.g and SCSFLR2031F05.g

In the column % Identity (Sb/At) corresponds to the amino acid sequence identity of sugarcane protein with Sorgum bicolor *and* Arabidopsis thaliana *homologs are shown.*
Only one sequence was found in Sorghum bicolor *with high similarity to sugarcane sequence and Arabidopsis's DEM, ROS1 and DMLE.*
**There are two NTH in* Arabidopsis thaliana*: NTH1 (Q9SIC4) and NTH2 (B9DFZ0).*
***There are two PCNA in* Arabidopsis thaliana*: PCNA1 (Q9M7Q7) and PCNA2. (Q9ZW35).*
****BER components that were identified sequence duplication in sugarcane genome.*

Table 1.
BER components from sugarcane.

Sugarcane BER components identified were compared with sequences belonging to *A. thaliana* and *Sorghum bicolor* (**Table 1**). Subsequently, these sequences were structurally and phylogenetically characterized; hence, their location on the pathway was set (**Table 1**). The following topics will address the particularities that were found relevant to BER and its specificities in sugarcane and monocots.

3. BER components—missing and differences

Sugarcane exhibits almost all components of the BER pathway, even though ROS1, DEM and DML3 were not identified as complete sequences. These DNA glycosylases, which play pivotal roles in epigenetic processes [27], have been well characterized [28–31] and were found in SUCEST-FUN as a single sequence without functional domains. Nevertheless, this result does not indicate a missing enzyme; epigenetic regulation is crucial, particularly in plants, and even more in polyploids organisms [32, 33]. Furthermore, the sugarcane database compiles numerous fragmented sequences that were not assembled and functionally annotated yet, as most data were from the transcriptome [25].

In contrast, differences were observed between the sequences of grasses analyzed (sugarcane and *S. bicolor*) when compared to dicotyledons, *A. thaliana* (**Table 2**). One of these differences, inconsistent with that observed in *A. thaliana,* was that the sequences of the *Poaceae* family present a second flap endonuclease protein 'FEN1B', which differs in size (as they are larger than the canonical Flap endonuclease 1 that receives the suffix A) and lacks the interaction sequence with PCNA. Notably, the sequences FEN1A and FEN1B are found at different *loci* and chromosomes of *S. bicolor*. Duplication in genes related to BER proteins was observed in AP endonucleases (ScARP1 and ScARP3) and MUMT (ScMUTM1 and ScMUTM2), which also reveal structural differences, as observed in FEN1A_CANA and FEN1B_CANA [34–36].

DNA ligase IV revealed certain differences regarding the domain disposition on the sequence (**Table 2**). Additionally, the sequences reveal variable identity (**Table 1**), thereby indicating high similarity within the grass plants. Notably, the BRCT domain is present in the sequences of *A. thaliana* and *S. bicolor*, but not in that of sugarcane. BRCT is a domain related to protein–protein interactions and is present in numerous proteins involved in DNA repair as well as cell cycle control [37–39]. Differences in domain disposition were also perceived in XRCC1, which displayed only one BRCT domain in the *Poaceae* family, whereas two BRCTs were found in the *A. thaliana* sequence. These differences could reflect variations in the protein role in DNA metabolism; these domains are essential because they comprise the activity and binding site of the enzyme.

4. BER's first step - base lesion recognition

BER is initiated by lesion-specific DNA glycosylases. The basic DNA glycosylase enzymatic process involves excision of the modified nucleobase from the DNA by catalyzing the hydrolysis of the N-glycosidic bond [40]. Regarding sugarcane, some of the BER's glycosylases were identified and characterized, suggesting the maintenance of the enzymes in *Saccharum spp.* as well as in conservation of the first step of BER pathway.

The DNA glycosylase OGG1 was identified in sugarcane and is called OGG1_CANA. This glycosylase as well as other sequences belonging to the *Poaceae* and the dicotyledonous, exhibit the conserved domain of the superfamily OGG1 [41].

Query	Accession	Protein domain	lenght (aa)
DML3_ARATH (O49498)	pfam15628	RRM_DME	1044
	cl23768	ENDO3c superfamily	
	cl21423	Perm-CXXC superfamily	
DME_ARATH (Q8LK56)	pfam15628	RRM_DME	1987
	cl23768	ENDO3c superfamily	
	pfam15629	Perm-CXXC	
	cl26620	Glutenin_hmw superfamily	
	cl34047	TonB superfamily	
ROS1_ARATH (Q9SJQ6)	pfam15628	RRM_DME	1393
	cl23768	ENDO3c superfamily	
	pfam15629	Perm-CXXC	
A0A1Z5R5E2_SORBI	pfam15628	RRM_DME	1878
	cl23768	ENDO3c superfamily	
	pfam15629	Perm-CXXC	
comp89337_c0_seq1	–	–	1469
FEN1_ARATH (O65251)	PF00867	N-domain	383
	PF00752	I-domain	
		Interaction with PCNA	
FEN1A_SORBI (C5YUK3)	PF00867	N-domain	380
	PF00752	I-domain	
		Interaction with PCNA	
FEN1B_SORBI (C5WU23)	PF00867	N-domain	428
	PF00752	I-domain	
FEN1A_CANA	PF00867	N-domain	379
	PF00752	I-domain	
		Interaction with PCNA	
FEN1B_CANA	PF00867	N-domain	413
	PF00752	I-domain	
DNLI4_ARATH (Q9LL84)	cl36689	dnl1 superfamily	1219
	cd17722	BRCT_DNA_ligase_IV_rpt1	
	cd17717	BRCT_DNA_ligase_IV_rpt2	
	cl31754	PTZ00121 superfamily	
A0A1Z5REU4_SORBI	cl36689	dnl1 superfamily	1281
	cd17722	BRCT_DNA_ligase_IV_rpt1	
	cl00038	BRCT superfamily	
	cl12940	DNA_ligase_IV superfamily	
DNLI4_CANA	cd07903	Adenylation_DNA_ligase_IV	572
	cl08424	OBF_DNA_ligase_family superfamily	
	pfam04675	DNA_ligase_A_N	

Query	Accession	Protein domain	lenght (aa)
XRCC1_ARATH	PRU00033	BRCT1	352
	PRU00033	BRCT2	
C5Z3V7_SORB	PS50172	BRCT	346
XRCC1_CANA	PS50172	BRCT	346

Table 2.
BER components with distinct features regarding protein domains in sugarcane.

This sequence reveals conservation of glutamine and phenylalanine residues (*Arabidopsis,* residues Q_{324} and F_{328}; sugarcane, Q_{378} and F_{382}) that are responsible for recognition of the damage base [42]. Moreover, site-directed mutagenesis assays in human OGG1 revealed that residues K_{249} and D_{268} (the sugarcane equivalent D_{334} and K_{315}) would also play an essential role in appropriate catalysis of DNA glycosylase [43, 44]. For MUTM, two sequences were identified in sugarcane: ScMUTM1 and ScMUTM2. Similar to OGG1_CANA, these sequences also retain essential residues for their enzymatic activity [36].

In *A. thaliana*, a homolog for Endonuclease III was identified and characterized, and termed as *Arabidopsis thaliana ENDONUCLEASE THREE HOMOLOG 1*(AtNTH1); it presented its enzymatic activity in relation to various substrates, thereby revealing its essential role in plant stress response [45]. A second endonuclease III homolog called AtNTH2, which was found together with AtNTH1 and AtARP in the *A. thaliana* chloroplast nucleus, demonstrating the occurrence of BER pathway in this organelle [46]. Considering grasses, a sequence that would refer to NTH2 remained unidentified. Phylogenetic analyses of this DNA glycosylase revealed duplication of sequences for organisms belonging to the group of dicots, but not for monocots.

Sugarcane NTH1, called NTH1_CANA, belongs to the Helix-hairpin-Helix (HHH) superfamily [47]. Furthermore, regarding *Escherichia coli*'s endonuclease III protein, the Helix-Hairpin-Helix domain has iron–sulfur binding sites [4Fe-4S] [48]. These sites comprised four conserved cysteines that would act on redox chemistry and DNA binding [49], and both motif and sites are conserved in the NTH1_CANA. Moreover, conservation of aspartic acid (D) at the active site, which is a residue preserved in other DNA glycosylases besides NTH1, such as UNG and MBD4L [50], was also evidenced in sugarcane.

Another glycosylase identified was UDG_CANA, which was conserved in the domain belonging to the UDG superfamily, more precisely concerning family-1 [51, 52]. Additionally, it conserved aspartic acid (D) as an active site [51]. It is known that the human UNG gene encodes two forms of the protein, one directed towards the mitochondria (UNG1) and another towards the nucleus (UNG2) [53]. The *A. thaliana* UNG (AtUNG) seems to be homologous to these two types of UNGs, being proven to act on mitochondrial DNA [54]. Most grass sequence annotations of computational prediction that directed the UNGs to both the nucleus and the mitochondria, raised the question whether there is only one UNG for both organelles in plants.

Ramiro-Merina et al. [55] demonstrated that *A. thaliana* encodes a monofunctional DNA glycosylase homologous to mammalian MBD4, known as MBD4-like or AtMBD4L. Nota et al. [56] indicated that the activation of AtMBD4L induces the expression of a late gene from the BER AtLIG1 pathway and reveals the mechanism by which it increases the plant's tolerance to oxidative stress. In relation to sugarcane, one fragment features the same domain and active site as AtMBD4L, implying a probable functional protein in *Saccharum spp.*

5. AP site removal—AP endonuclease role in sugarcane

AP endonuclease is an essential enzyme for BER pathway as this enzyme identifies and process AP (apurinic/apyrimidinic) site [57]. These AP sites may be a result of the action of DNA glycosylases or it may be spontaneously generated. Unrepaired AP sites can lead to mutations during semiconservative replication, which indicates the importance of the role of AP endonuclease in maintenance of the genetic code [58].

In *A. thaliana*, three AP endonucleases are homologous to APE1 (HUMAN AP ENDONUCLEASE 1), namely AtAPE1L, AtAPE2, and AtARP [59]. Each of these presents their specifics based on enzymatic activity, regulation, and sub-cellular localization. For sugarcane, two sequences were identified, and their three-dimensional structures were inferred: ScARP1 and ScARP3 [35]. By examining the sequences of ScARPs (1 and 3), we can observe the conservation of essential sites for the catalysis and binding of metals (enzymatic cofactors) [34].

ScARP1 has greater similarity with AtARP (60%), whereas AtAPE1L and AtAPE2 reveal a correspondence below 50%. These values may indicate diversity in structure, amino acid composition, and perhaps function. ScARP3 reveals a divergence compared with ScARP1. ScARP3 is closer to AtARP, presenting an even higher percentage of identity (75%). Maíra et al. [35] demonstrated that the sequence ScARP3 would be closer to the group of dicotyledonous plants, whereas ScARP1 would be included within the monocots, more precisely together with representatives of the *Poaceae* family.

Medeiros et al. [34] purified ScARP1 and verified the enzymatic activity of this sugarcane enzyme against several substrates. This study found the capability of ScARP1 to process AP sites; however, other enzymatic activities (exonuclease, phosphatase, and 3′-phosphodiesterase) were not confirmed. The AP endonuclease activity complementation assay in extracts of *A. thaliana* demonstrated that ScARP1 was capable of complementing around 40% of the activity of AtARP from $arp^{-/-}$ mutant plant extracts [34].

6. Flap endonuclease (FEN1)—BER's long-patch in sugarcane

FEN1 is a structure-specific nuclease that can remove flap structures and is involved in different DNA metabolic pathways, including DNA replication, DNA repair, apoptotic DNA degradation, and maintenance of telomere stability [60]. In case of BER, FEN1 in complex with proliferating cell nuclear antigen (PCNA) plays a pivotal role in the long patch as it removes a short flap structure generated by Pol β activity [61].

Regarding plants, it is known that two FEN1 counterparts were identified in rice (*Oryza sativa*, OsFEN1a, OsFEN1b). Functional complementation assays revealed that only OsFEN1a would be able to complement the *fen1/rad27* mutant in yeast, suggesting that these two genes may be functionally distinct [62]. In addition, OsFEN1a, expressed in *Escherichia coli*, presents flap-endonuclease and 5′-exonuclease activity [63]. In *A. thaliana*, only one FEN1 homolog, namely SAV6, was identified [63]. Biochemical characterization of SAV6 protein (also called FEN1) revealed that, unlike animal FEN1, the SAV6 protein has flap-endonuclease and gap endonuclease activity but does not reveal 5′ exonuclease activity; however, similar to human FEN1 (hFEN1), SAV6 is also necessary for maintaining the genome integrity and responding to plant DNA damage [64].

As observed in *O. sativa*, sugarcane has two FEN1 sequences, namely FEN1A_CANA and FEN1B_CANA. Considering the structure of the Flap endonuclease, it is known that

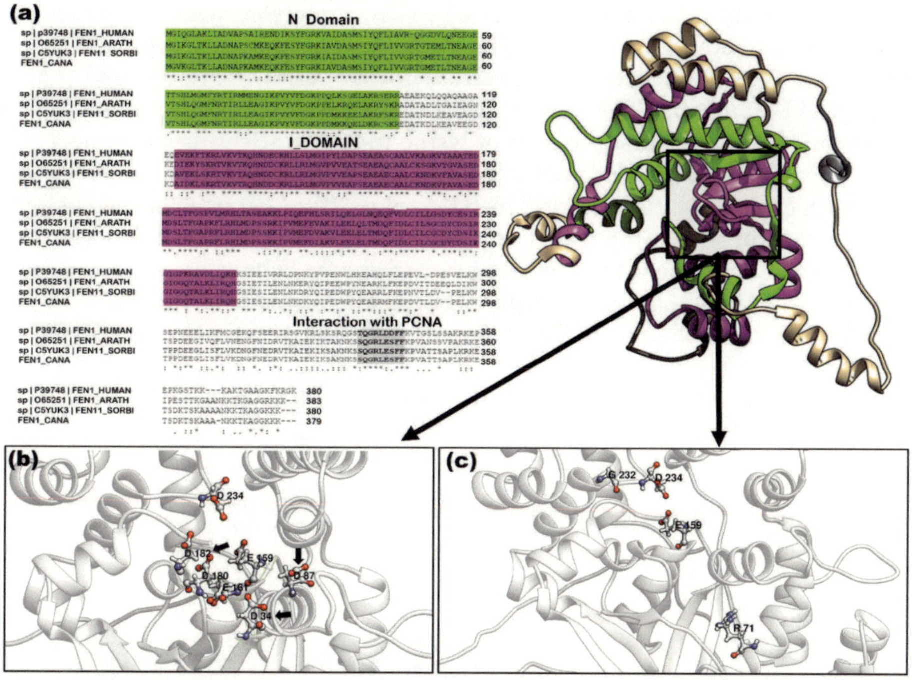

Figure 1.
The proposed model for FEN1 of sugarcane. (a) It was represented the alignment obtained using Clustal omega for FEN1 sequences of Homo sapiens, Arabidopsis thaliana, Sorghum bicolor *and sugarcane. The colors in the alignment and in the model, correspond to the N-terminal region (green), internal region (I) (purple) and the segment that interacted with PCNA (gray). Metal-binding sites (b) and DNA binding sites (c) are highlighted. The black arrows in (b) indicate the probable active site of the enzyme.*

human FEN1 comprises the N-terminal domain and the intermediate (Domain I) in addition to a C-terminal region, which is important for the interaction of FEN1 with other proteins, such as PCNA and WRN (**Figure 1**) [65, 66]. The FEN1A_CANA sequence preserves the domains described previously; however, the FEN1B_CANA and other *Poaceae* similar sequences analyzed do not possess the binding domain for PCNA in its C-terminal region, which may affect its mechanism of action in the plant cell.

Considering the protein structure, FEN1 is a nuclease that features two regions: the N-terminal region and I-region [67]. The alignment of FEN1 from *H. sapiens*, *A. thaliana* and *S. bicolor* as well as FEN1A's sugarcane ascertained the conservation of these regions (**Figure 1**). Notably, the region of interaction with the PCNA (**Figure 1a**) that is in a loop, in that way, more exposed and facilitating its possible interaction with PCNA. The sugarcane FEN1A model presents the conservation of metal-binding sites (Mg^{+2} ion; **Figures 1b** and **c**). The residues D_{34}, D_{87}, and D_{182}, considering the equivalent residues in human FEN1 [68], may be responsible for the catalytic activity of the enzyme (**Figure 1b** and **c**).

7. PCNA role in plants

Studies on PCNA have revealed that it plays a crucial role in DNA replication as well as in DNA repair, cell cycle regulation and apoptosis [69–71]. In *A. thaliana*, two PCNAs, AtPCNA1 and AtPCNA2, are present, which differ from each other in eight amino acids, in addition to the fact that AtPCNA2 has an extra residue in the protein length [72]. Of these eight different amino acids, four are identical to the residues found in *Brassica napus* and human PCNAs [73].

Considering the difference between AtPCNAs, Anderson et al. [72] demonstrated that co-expression of POLH (DNA polymerase eta - Pol η) and AtPCNA2 (and not AtPCNA1) was necessary to restore normal resistance to UV radiation in the yeast RAD30 mutant. The difference was in lysine (K) 201 present in AtPCNA1, which would inhibit the ubiquination of lysine 164, thus affecting its connection with Pol η and not being able to act on trans-lesion synthesis (TLS) and restore the progression of the replication fork. The lysine at position 201 of AtPCNA1 belonged to the group comprising amino acids with electrically charged side chains. In the case of K, this could be endowed with a positive charge, whereas the corresponding one at AtPCNA2 would be an asparagine (N) that belonged to the group of amino acids with polar side chains without being loaded. In PCNA_CANA, the corresponding residue in question would be a glutamine that concerns the same group as N, which leads to the conclusion that sugarcane PCNA would be closer to AtPCNA2 than AtPCNA1 and could, as such, act in the TLS.

The three-dimensional model of sugarcane's PCNA is revealed as homotrimeric architecture in the form of a ring, comprising three identical chains of PCNA, as indicated by different colors in the sugarcane PCNA model (**Figure 2**).

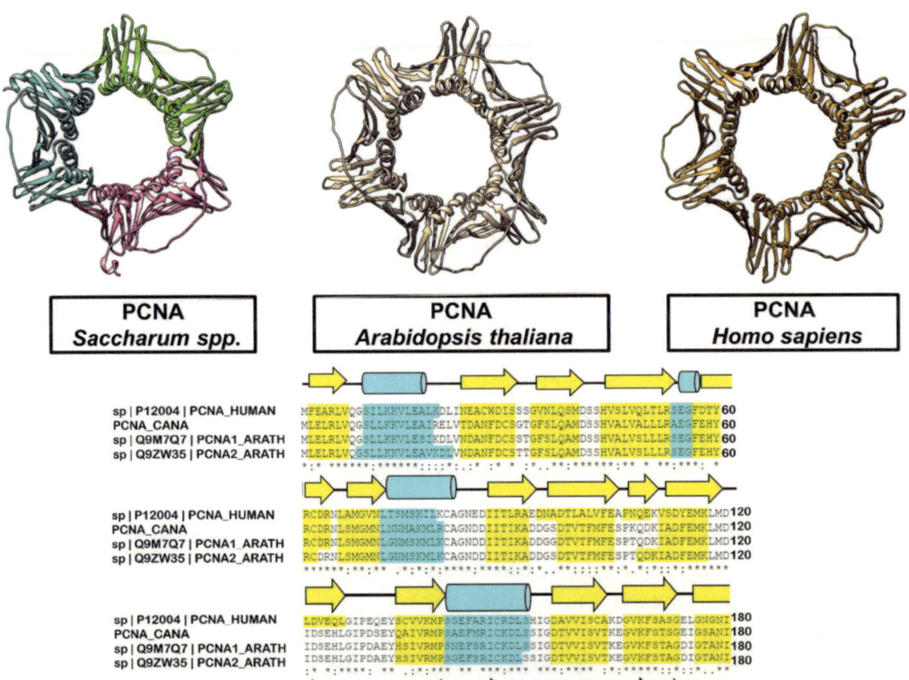

Figure 2.
Three-dimensional model and protein sequences of plant and human PCNAs. The 3D models are depicted above the alignment presenting conservation of the structure in ring-shaped homotrimeric architectures. The model of the putative sugarcane PCNA; the structures highlighted in blue, green and pink are individual chains of PCNA that together compose the homotrimeric ring. Below the models, the corresponding alignment of the PCNA sequences of Arabidopsis thaliana, Homo sapiens *and* Saccharum spp., *highlighting the secondary structures (yellow arrow, beta sheet; blue cylinder, alfa helix) is presented.*

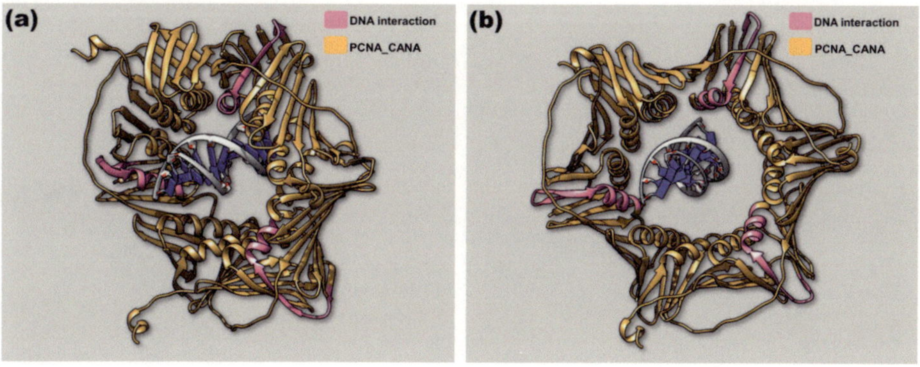

Figure 3.
Models proposed for sugarcane PCNA associated with DNA. The region of the PCNA that interacts with the DNA, facing the inside of the ring of the homotrimeric structure, is depicted in pink. The double strand that constitutes a helix of predominantly blue color represents the three-dimensional structure of DNA. (a) View of the sugarcane PCNA model (in orange) interacting with the DNA (structure in blue and white) seen from the side. (b) Frontal view of the model.

Additionally, this model exhibits sequence and structural similarity with other PCNAs, as observed in **Figure 2**. Compared to *A. thaliana* and *H. sapiens* PCNAs, sugarcane's PCNA overlapped its secondary structure. Structure conservation is also observed in the PCNA models (**Figure 2**), which display the same homotrimeric ring predicted for the sugarcane model.

PCNA is generally called a sliding clamp, since it was predicted that the double strand of DNA would pass through the opening of the PCNA ring and would serve as an anchoring platform for several proteins involved in DNA metabolism [74]. The attainable preservation of this function was verified in the *Saccharum spp*. model. The 5L7C crystal [75], which is a model of human PCNA, was used for comparison with PCNA's sugarcane. By aligning the crystal with the model (root mean square deviation, RMSD = 0.653 Å), it was possible to ascertain the probable conservation of its interaction with the DNA (depicted in pink). Therefore, we identified a DNA binding site in the model, which faces the interior of the ring orifice, where the double strand of DNA should pass (depicted in blue) (**Figure 3**).

8. Sugarcane protein models—conservation throughout plants

PCNA and FEN1A were proteins identified in sugarcane, which were presumed to interact with each other [76]. This is due to PCNA interaction sequence detected in the N-terminal segment of FEN1A. To verify the truthfulness of this interaction, three-dimensional models were created for PCNA and FEN1A sugarcane proteins. These models were assessed for the conservation of secondary structure, active sites, and residue interactions with the substrate. Based on this analysis, the role of these sugarcane proteins can be established.

PCNA, as previously mentioned, would serve as a scaffold, and moreover, various functions can be performed ranging from DNA methylation to base excision repair. Thus, using the IUL1 crystal that comprises the human PCNA associated with FEN1 [77], the possibility of the sugarcane's predicted models of these proteins that may interact with each other was verified. This result demonstrates that FEN1 of sugarcane is associated with the homotrimeric ring of PCNA (**Figure 4**). The sequence of interaction with PCNA differs, revealing that this sequence is in the interface of PCNA and FEN1 interaction.

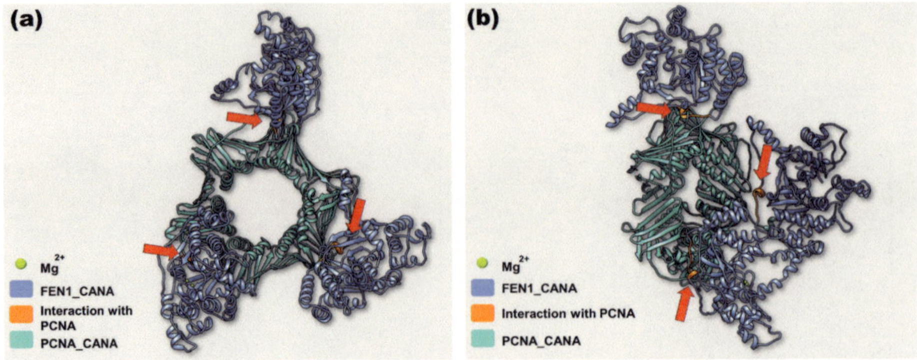

Figure 4.
Proposed complex of sugarcane PCNA and FEN1. (a) Lateral view of the complex. (b) Frontal view of the complex. FEN1 models are in blue, PCNA in green and Mg^{+2} ions are highlighted in yellow. The sequence of FEN1 that interacts with PCNA is highlighted in orange and is indicated with red arrows.

9. BER pathway—evolutionary analysis in the grass outlook

Overall, the phylogenetic analyses revealed differences in the presence or absence of duplication of BER pathway components. In few cases, duplication was observed in dicotyledons and not in monocotyledons, for example, NTH, PCNA and DNA ligase 1. Herein, structural difference was noted (size, presence or absence of certain conserved domains), indicating diverse DNA repair mechanisms between plants.

Singh et al. [78] compared the plant genomes available at that time, thus aiming to compare the genes involved in DNA repair and recombination. They found that FEN1, in the genome of monocotyledons (corn, rice, *S. bicolor*, and *Brachypodium distachyon*) presented two copies and that such copies would not be products of intra-genomic duplication. In particular, these copies were subtypes of FEN1, FEN1A and FEN1B. Singh et al. [78] also identified one copy of FEN1 in dicots, namely *A. thaliana*, *Medicago truncatula*, *Vitis venifera*, and *Papaver somniferum*; however, *Glycine max* presented two copies of FEN1; in such case, these copies were products of intra-genomic duplication.

A new analysis regarding FEN1 in plants, particularly sugarcane, was conducted. It was discovered that FEN1B was only found within *Poales*, specifically *Panicoideae* (**Figure 5**). Important crops such as *Oryza sativa*, *Zea mays* and *S. bicolor* display FEN1B as well as FEN1A. Evolutionary analyses revealed that FEN1A and FEN1B had distinct assembly. Moreover, the flap endonucleases (FEN1A and FEN1B) of the same species were not located at the same branch in the phylogenetic tree. Nonetheless, FEN1 was duplicated in some eudicot groups, as in *Noccaea caerulescens* and *Nicotiana tabacum*; however, these sequences have all the regions required for a functional FEN1.

Although the absence of region may compromise the enzymatic activity of FEN1B, the other residues, domains and active sites were conserved. These findings raise questions regarding the maintenance of FEN1B in the genome of these organisms, its functions and its role in BER.

Maíra et al. [35] proposed that a whole genome duplication event (WGD) would be related to the duplication observed in the AP endonuclease sequence in the grasses group; however, further studies indicate that duplications are present in other plant groups in addition to *Poaceae*. The BER's duplication genes do not cover all the components of this pathway; on the contrary, a few sequences—ARP, MUTM and FEN1—could be set as duplications. Issues regarding the maintenance

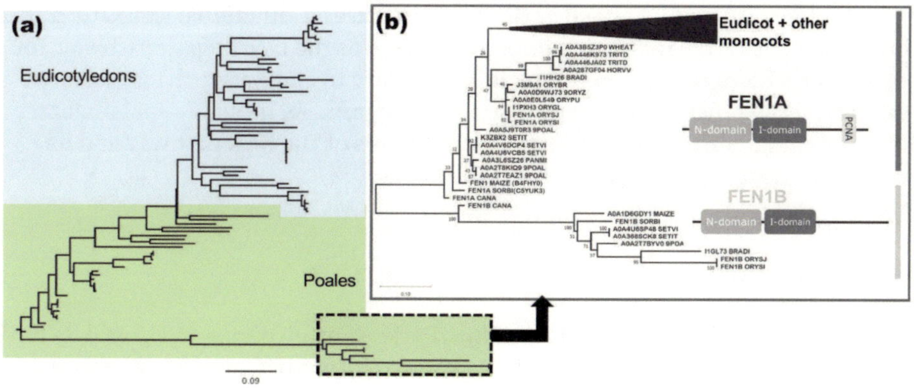

Figure 5.
FEN1 evolutionary analysis by maximum likelihood method. The evolutionary history was inferred using the maximum likelihood method and JTT matrix-based model conducted in MEGA X. the percentage of trees in which the associated taxa clustered together is presented next to the branches. (a) Phylogenetic tree comprising plant FEN1 sequences; green color represents the Poales group and blue color represents the eudicotyledons group. (b) Phylogenetic tree focus on Poales group, in which FEN1A and FEN1B clusters are displayed on distinct branches. The FEN1A and FEN1B domains are displayed next to their respective clusters.

of these sequences in plant genomes, particularly sugarcane, need to be responded to essentially comprehend the evolutionary aspect of the BER pathway in monocots.

Notably, the fate of the vast majority of duplicate genes resulting from segmental duplication includes the nonfunctionalization of a member of the pair [79, 80], which should occur within a few million years in the absence of any intrinsic advantage of duplicate copying [80, 81]. Specifically, plant genomes, on average, reveal 65% of their annotated genes that are duplicated [82]. Most of these copies are derived from ancient WGD events in the terrestrial plant lineage [82]. Li et al. [83] investigated the fate of duplicate genes from 40 different species of flowering plants; of these, all species experienced at least one or more WGD events throughout their evolutionary history. The loss of genes was observed immediately after genome duplication, so that the genes quickly returned to the state of a single copy [83]; however, some of these genes have preserved their state of multiple copies. Such genes belong to families of genes involved in the response to biotic and abiotic stress, and are therefore important for the adaptation of the plant to the environment. Thus, it is possible to correlate the duplication and retention of these copies with an adaptive advantage such that genes can confer to the plant, allowing it to act more efficiently in response to environmental variations. DNA repair genes are linked to this hypothesis, since they are necessary to maintain the stability of the genome and preserve genetic information. In addition to the fact that several of these genes, already described in this chapter, act in other processes of adaptive importance such as response to oxidative stress.

10. Conclusions

In sugarcane as well as in other plants, except for the plant models, few studies have focused on the characterization and structural analysis of individual components of metabolic pathways. Moreover, it should be considered that the traditional breeding strategy lags behind the demand for commercial needs due to insufficient knowledge on characteristics related to stress tolerance, inefficient selection techniques and low genetic variation and fertility. The evident deficiency of biotechnology will be supplemented with studies aimed at the biochemical and functional characterization of important pathways and their components, such as the DNA repair pathway, for instance, the BER pathway. Therefore, it is necessary to emphasize the importance of this chapter in other plant species apart from sugarcane, provide supplementary information, and raise questions on the components of the BER pathway and its evolutionary issue regarding monocots and dicots.

Acknowledgements

The authors wish to thank to Coordenação de Aperfeiçoamento Pessoal de Nível Superior-CAPES, Conselho Nacional de Desenvolvimento Científico e Tecnológico-CNPq, and Ministério de Ciência, Tecnologia, Inovação (MCTI) for financial support.

Conflict of interest

The authors declare no conflict of interest.

Author details

Nathalia Maíra Cabral de Medeiros and Katia Castanho Scortecci*
Departamento de Biologia Celular e Genética – Centro de Biociencias, Universidade Federal do Rio Grande do Norte, Brazil

*Address all correspondence to: kacscort@yahoo.com

IntechOpen

© 2021 The Author(s). Licensee IntechOpen. This chapter is distributed under the terms of the Creative Commons Attribution License (http://creativecommons.org/licenses/by/3.0), which permits unrestricted use, distribution, and reproduction in any medium, provided the original work is properly cited.

References

[1] Wallace SS. Base Excision Repair: A Critical Player in Many Games. DNA Repair. 2014;19:14-26. DOI: 10.1016/j.dnarep.2014.03.030.

[2] Lindahl T, Wood RD. Quality Control by DNA Repair. Science. 1999;286:1897-1905. DOI: 10.1126/science.286.5446.1897.

[3] Krokan HE, Bjørås M. Base excision repair. Cold Spring Harb Perspect Biol. 2013;5:a012583. DOI: 10.1101/cshperspect.a012583.

[4] Drohat AC, Coey CT. Role of Base Excision "repair" Enzymes in Erasing Epigenetic Marks from DNA. Chemical Reviews. 2016;166:12711-12729. DOI: 10.1021 acs.chemrev.6b00191.

[5] Kohli RM, Zhang Y. TET Enzymes, TDG and the Dynamics of DNA Demethylation. Nature. 2013;502:472-9 DOI: 10.1038/nature12750.

[6] Fleming AM, Ding Y, Burrows CJ. Oxidative DNA damage is epigenetic by regulating gene transcription via base excision repair. Proc Natl Acad Sci U S A. 2017;114:2604-2609. DOI:10.1073/pnas.1619809114.

[7] Bellacosa A, Drohat AC. Role of base excision repair in maintaining the genetic and epigenetic integrity of CpG sites. DNA Repair (Amst). 2015; 32:33-42. DOI: 10.1016/j.dnarep.2015.04.011.

[8] Roldan-Arjona T, Ariza RR, Córdoba-Cañero D. DNA base excision repair in plants: an unfolding story with familiar and novel characters. Frontiers in plant science. 2019;10:1055. DOI: 10.3389/fpls.2019.01055.

[9] Kimura S, Sakaguchi K. DNA repair in plants. Chemical Reviews. 2006;106:753-766. DOI: 10.1021/cr040482n.

[10] Britt AB. DNA damage and repair in plants. Annu Rev Plant Physiol Plant Mol Biol. 1996;47:75-100. DOI: 10.1146/annurev.arplant.47.1.75.

[11] Manova V, Gruszka D. DNA damage and repair in plants – From models to crops. Frontiers in Plant Science. 2015;6:885. DOI: 10.3389/fpls.2015.00885.

[12] Sarwar H, Sarwar MF, Sarwar M, Qadri NA, Moghal S. The importance of cereals (Poaceae: Gramineae) nutrition in human health: A review. Journal of cereals and oilseeds. 2013;4:32-35 .DOI: 10.5897/jco12.023.

[13] FAO. Food and Agriculture Data [Internet].2018. Available from: http://www.fao.org/faostat/en/#home [Accessed: 2020-06-19].

[14] Thirugnanasambandam PP, Hoang N V., Henry RJ. The challenge of analyzing the sugarcane genome. Frontiers in Plant Science. 2018;9:616. DOI: 10.3389/fpls.2018.00616.

[15] Souza GM, Berges H, Bocs S, Casu R, D'Hont A, Ferreira JE, Henry R, Ming R, Potier B, Van Sluys MA, Vincentz M, Paterson AH. The Sugarcane Genome Challenge: Strategies for Sequencing a Highly Complex Genome. Trop Plant Biol. 2011;4:145-156. DOI: 10.1007/s12042-011-9079-0.

[16] Dal-Bianco M, Carneiro MS, Hotta CT, Chapola RG, Hoffmann HP, Garcia AAF, Souza GM. Sugarcane improvement: How far can we go? Current Opinion in Biotechnology. 2012;23:265-270. DOI: 10.1016/j.copbio.2011.09.002.

[17] Lima WC, Medina-Silva R, Galhardo RS, Menck CFM. Distribution of DNA repair-related ESTs in sugarcane. Genet Mol Biol. 2001;24:141-146. DOI: 10.1590/S1415-47572001000100019.

[18] Costa RMA, Lima WC, Vogel CIG, Berra CM, Luche DD, Medina-Silva R, Galhardo RS, Menck CFM, Oliveira VR. DNA repair-related genes in sugarcane expressed sequence tags (ESTs). Genet Mol Biol. 2001;24:131-140. DOI: 10.1590/S1415-47572001000100018.

[19] Agnez-Lima LF, de Medeiros SRB, Maggi BS, Quaresma GAS. Base excision repair in sugarcane. Genet Mol Biol. 2001;24:123-129. DOI: 10.1590/S1415-47572001000100017.

[20] Okura VK, de Souza RSC, de Siqueira Tada SF, Arruda P. BAC-pool sequencing and assembly of 19 Mb of the complex sugarcane genome. Front Plant Sci. 2016;7:342. DOI: 10.3389/fpls.2016.00342.

[21] De Mendonça VM, Del Bem LE, Van Sluys MA, De Setta N, Kitajima JP, Cruz GMQ, Sforça DA, De Souza AP, Ferreira PCG, Grativol C, Cardoso-Silva CB, Vicentini R, Vincentz M. Analysis of Three Sugarcane Homo/Homeologous Regions Suggests Independent Polyploidization Events of *Saccharum officinarum* and *Saccharum spontaneum*. Genome Biol Evol. 2017;9:266-278, DOI: 10.1093/gbe/evw293.

[22] Grativol C, Regulski M, Bertalan M, McCombie WR, Da Silva FR, Zerlotini Neto A, Vicentini R, Farinelli L, Hemerly AS, Martienssen RA, Ferreira PCG. Sugarcane genome sequencing by methylation filtration provides tools for genomic research in the genus Saccharum. Plant J. 2014;79:162-172. DOI: 10.1111/tpj.12539.

[23] Riaño-Pachón DM, Mattiello L. Draft genome sequencing of the sugarcane hybrid SP80-3280. F1000Research. 2017;6:861. DOI: 10.12688/f1000research.11859.2

[24] Piperidis N, D'Hont A. Sugarcane genome architecture decrypted with chromosome-specific oligo probes. Plant J. 2020;103:2039-2051. DOI: 10.1111/tpj.14881.

[25] Vettore AL, da Silva FR, Kemper EL, Souza GM, da Silva AM, Ferro MIT, Henrique-Silva F, Giglioti ÉA, Lemos MVF, Coutinho LL, Nobrega MP, Carrer H, França SC, Bacci M, Goldman MHS, Gomes SL, Nunes LR, Camargo LEA, Siqueira WJ, Van Sluys MA, Thiemann OH, Kuramae EE, Santelli R V., Marino CL, Targon MLPN, Ferro JA, Silveira HCS, Marini DC, Lemos EGM, Monteiro-Vitorello CB, Tambor JHM, Carraro DM, Roberto PG, Martins VG, Goldman GH, de Oliveira RC, Truffi D, Colombo CA, Rossi M, de Araujo PG, Sculaccio SA, Angella A, Lima MMA, de Rosa VE, Siviero F, Coscrato VE, Machado MA, Grivet L, Di Mauro SMZ, Nobrega FG, Menck CFM, Braga MDV, Telles GP, Cara FAA, Pedrosa G, Meidanis J, Arruda P. Analysis and functional annotation of an expressed sequence tag collection for tropical crop sugarcane. Genome Res. 2003;13:2725-2735. DOI:10.1101/gr.1532103.

[26] Yilmaz A, Nishiyama MY, Fuentes BG, Souza GM, Janies D, Gray J, Grotewold E. GRASSIUS: A platform for comparative regulatory genomics across the grasses. Plant Physiol. 2009;149:171-180. DOI:10.1104/pp.108.128579.

[27] Liu R, Lang Z. The mechanism and function of active DNA demethylation in plants. Journal of Integrative Plant Biology. 2020;62:148-159. DOI:10.1111/jipb.12879.

[28] Agius F, Kapoor A, Zhu JK. Role of the Arabidopsis DNA glycosylase/lyase ROS1 in active DNA demethylation. Proc Natl Acad Sci U S A. 2006;103:11796-11801. DOI:10.1073/pnas.0603563103.

[29] Tang K, Lang Z, Zhang H, Zhu JK. The DNA demethylase ROS1 targets genomic regions with distinct chromatin

modifications. Nat Plants. 2016;2:1-10. DOI:10.1038/nplants.2016.169.

[30] Morales-Ruiz T, Ortega-Galisteo AP, Ponferrada-Marín MI, Martínez-Macías MI, Ariza RR, Roldán-Arjona T. DEMETER and REPRESSOR OF SILENCING 1 encode 5-methylcytosine DNA glycosylases. Proc Natl Acad Sci U S A. 2006;103:6853-6858. DOI:10.1073/pnas.0601109103.

[31] Liu R, How-Kit A, Stammitti L, Teyssier E, Rolin D, Mortain-Bertrand A, Halle S, Liu M, Kong J, Wu C, Degraeve-Guibault C, Chapman NH, Maucourt M, Hodgman TC, Tost J, Bouzayen M, Hong Y, Seymour GB, Giovannoni JJ, Gallusci P. A DEMETER-like DNA demethylase governs tomato fruit ripening. Proc Natl Acad Sci U S A. 2015;112:10804-10809. DOI:10.1073/pnas.1503362112.

[32] Ding M, Chen ZJ. Epigenetic perspectives on the evolution and domestication of polyploid plant and crops. Current Opinion in Plant Biology. 2018;42: 37-48. DOI:10.1016/j.pbi.2018.02.003.

[33] Mason AS, Wendel JF. Homoeologous Exchanges, Segmental Allopolyploidy, and Polyploid Genome Evolution. Frontiers in Genetics. 2020;11:1014. DOI:10.3389/fgene.2020.01014.

[34] Medeiros NMC, Córdoba-Cañero D, García-Gil CB, Ariza RR, Roldán-Arjona T, Scortecci KC. Characterization of an AP endonuclease from sugarcane – ScARP1. Biochem Biophys Res Commun. 2019;514:926-932. DOI: 10.1016/j.bbrc.2019.04.156

[35] Maira N, Torres TM, de Oliveira AL, de Medeiros SRB, Agnez-Lima LF, Lima JPMS, Scortecci KC. Identification, characterisation and molecular modelling of two AP endonucleases from base excision repair pathway in sugarcane provide insights on the early evolution of green plants. Plant Biol. 2014;16:622-631. DOI:10.1111/plb.12083.

[36] Scortecci KC, Lima AFO, Carvalho FM, Silva UB, Agnez-Lima LF, de Medeiros SRB. A characterization of a MutM/Fpg ortholog in sugarcane-A monocot plant. Biochem Biophys Res Commun. 2007;361:1054-1060. DOI: 10.1016/j.bbrc.2007.07.134.

[37] Callebaut I, Mornon JP. From BRCA1 to RAP1: A widespread BRCT module closely associated with DNA repair. FEBS Lett. 1997;400:25-30. DOI: 10.1016./S0014-5793(96)01312-9.

[38] Reubens MC, Rozenzhak S, Russell P. Multi-BRCT Domain Protein Brc1 Links Rhp18/Rad18 and γH2A To Maintain Genome Stability during S Phase. Mol Cell Biol. 2017;37:e00260-17. DOI: 10.1128/MCB.00260-17.

[39] Polo LM, Xu Y, Hornyak P, Garces F, Zeng Z, Hailstone R, Matthews SJ, Caldecott KW, Oliver AW, Pearl LH. Efficient Single-Strand Break Repair Requires Binding to Both Poly(ADP-Ribose) and DNA by the Central BRCT Domain of XRCC1. Cell Rep. 2019;26:573-581. DOI:10.1016/j.celrep.2018.12.082.

[40] Krokan HE, Standal R, Slupphaug G. DNA glycosylases in the base excision repair of DNA. Biochemical Journal. 1997;325:1-16. DOI:10.1042/bj3250001.

[41] Dany AL, Tissier A. A functional OGG1 homologue from *Arabidopsis thaliana*. Mol Gen Genet. 2001; 265:293-301. DOI:10.1007/s004380000414.

[42] García-Ortiz MV, Ariza RR, Roldán-Arjona T. An OGG1 orthologue encoding a functional 8-oxoguanine DNA glycosylase/lyase in *Arabidopsis thaliana*. Plant Mol Biol. 2001;47:795-804. DOI: 10.1023/A:1013644026132.

[43] Nash HM, Lu R, Lane WS, Verdine GL. The critical active-site amine of the human 8-oxoguanine DNA glycosylase, hOgg1: Direct identification, ablation and chemical reconstitution. Chem Biol. 1997;4:693-702. DOI: 10.1016/s1074-5521(97)90225-8.

[44] Bjørås M, Seeberg E, Luna L, Pearl LH, Barrett TE. Reciprocal "flipping" underlies substrate recognition and catalytic activation by the human 8-oxo-guanine DNA glycosylase. J Mol Biol. 2002;317:171-177. DOI:10.1006/jmbi.2002.5400.

[45] Roldan-Arjona T, Garcia-Ortiz M V., Ruiz-Rubio M, Ariza RR. cDNA cloning, expression and functional characterization of an *Arabidopsis thaliana* homologue of the *Escherichia coli* DNA repair enzyme endonuclease IIIcDNA cloning, expression and functional characterization of an *Arabidopsis thaliana* homologue of the *Escherichia coli* DNA repair enzyme endonuclease III. Plant Mol Biol. 2000;44:43-52. DOI: 10.1023/A:1006429114451.

[46] Gutman BL, Niyogi KK. Evidence for base excision repair of oxidative DNA damage in chloroplasts of *Arabidopsis thaliana*evidence for base excision repair of oxidative DNA damage in chloroplasts of *Arabidopsis thaliana*. J Biol Chem. 2009;284:17006-17012. DOI: 10.1074/jbc.M109.008342.

[47] Choi Y, Gehring M, Johnson L, Hannon M, Harada JJ, Goldberg RB, Jacobsen SE, Fischer RL. DEMETER, a DNA glycosylase domain protein, is required for endosperm gene imprinting and seed viability in Arabidopsis. Cell. 2002;110:33-42. DOI:10.1016/S0092-8674(02)00807-3.

[48] Kuo CF, McRee DE, Cunningham RP, Tainer JA. Crystallization and crystallographic characterization of the iron-sulfur-containing DNA-repair enzyme endonuclease III from *Escherichia coli*. J Mol Biol. 1992;227:347-351. DOI:10.1016/0022-2836(92)90703-M.

[49] Thayer MM, Ahern H, Xing D, Cunningham RP, Tainer JA. Novel DNA binding motifs in the DNA repair enzyme endonuclease III crystal structure. EMBO J. 1995;14:4108-4120. DOI: 10.1002/j.1460-2075.1995.tb00083.x.

[50] Choi Y, Gehring M, Johnson L, Hannon M, Harada JJ, Goldberg RB, Jacobsen SE, Fischer RL. DEMETER, a DNA glycosylase domain protein, is required for endosperm gene imprinting and seed viability in Arabidopsis. Cell. 2002;110:33-42. DOI: 10.1016/S0092-8674(02)00807-3.

[51] Pearl LH. Structure and function in the uracil-DNA glycosylase superfamily. Mutat Res - DNA Repair. 2000;460:165-181. DOI:10.1016/S0921-8777(00)00025-2.

[52] Lee HW, Dominy BN, Cao W. New family of deamination repair enzymes in Uracil-DNA glycosylase superfamily. J Biol Chem. 2011;286: 31282-31287. DOI:10.1074/jbc.M111.249524.

[53] Krokan HE, Kavli B, Sarno A, Slupphaug G. Enzymology of Genomic Uracil Repair. Genomic Uracil. 2018; 89-126. DOI:10.1142/9789813233508_0004.

[54] Boesch P, Ibrahim N, Paulus F, Cosset A, Tarasenko V, Dietrich A. Plant mitochondria possess a short-patch base excision DNA repair pathway. Nucleic Acids Res. 2009; 37:5690-5700. DOI:10.1093/nar/gkp606.

[55] Ramiro-Merina Á, Ariza RR, Roldán-Arjona T. Molecular characterization of a putative plant homolog of MBD4 DNA glycosylase. DNA Repair (Amst). 2013;12:890-898. DOI:10.1016/j.dnarep.2013.08.002.

[56] Nota F, Cambiagno DA, Ribone P, Alvarez ME. Expression and function of AtMBD4L, the single gene encoding the nuclear DNA glycosylase MBD4L in Arabidopsis. Plant Sci. 2015;235:122-129. DOI:10.1016/j.plantsci.2015.03.011.

[57] Wiederhold L, Leppard JB, Kedar P, Karimi-Busheri F, Rasouli-Nia A, Weinfeld M, Tomkinson AE, Izumi T, Prasad R, Wilson SH, Mitra S, Hazra TK. AP endonuclease-independent DNA base excision repair in human cells. Mol Cell. 2004;15:209-220. DOI:10.1016/j.molcel.2004.06.003.

[58] Talpaert-Borlè M. Formation, detection and repair of AP sites. Mutat Res - Fundam Mol Mech Mutagen. 1987;181:45-56. DOI:10.1074/jbc.M111.249524.

[59] Murphy TM, Belmonte M, Shu S, Britt AB, Hatteroth J. Requirement for abasic endonuclease gene homologues in Arabidopsis seed development. PLoS One. 2009;4:e4297. DOI: 10.1371/journal.pone.0004297.

[60] Tsutakawa SE, Classen S, Chapados BR, Arvai AS, Finger LD, Guenther G, Tomlinson CG, Thompson P, Sarker AH, Shen B, Cooper PK, Grasby JA, Tainer JA. Human flap endonuclease structures, DNA double-base flipping, and a unified understanding of the FEN1 superfamily. Cell. 2011;145:198-211. DOI:10.1016/j.cell.2011.03.004.

[61] Lin Y, Beard WA, Shock DD, Prasad R, Hou EW, Wilson SH. DNA polymerase β and flap endonuclease 1 enzymatic specificities sustain DNA synthesis for long patch base excision repair. J Biol Chem. 2005;280:3665-3674. DOI:10.1074/jbc.M412922200.

[62] Kimura S, Furukawa T, Kasai N, Mori Y, Kitamoto HK, Sugawara F, Hashimoto J, Sakaguchi K. Functional characterization of two flap endonuclease-1 homologues in rice. Gene. 2003;314:63-71. DOI:10.1016/S0378-1119(03)00694-2.

[63] Kimura S, Ueda T, Hatanaka M, Takenouchi M, Hashimoto J, Sakaguchi K. Plant homologue of flap endonuclease-1: Molecular cloning, characterization, and evidence of expression in meristematic tissues. Plant Mol Biol. 2000;42:415-427. DOI:10.1023/A:1006349511964.

[64] Zhang Y, Wen C, Liu S, Zheng L, Shen B, Tao Y. Shade avoidance 6 encodes an arabidopsis flap endonuclease required for maintenance of genome integrity and development. Nucleic Acids Res. 2016;44:1271-1284. DOI:10.1093/nar/gkv1474.

[65] Karanja KK, Livingston DM. C-terminal flap endonuclease (rad27) mutations: Lethal interactions with a DNA ligase I mutation (cdc9-p) and suppression by proliferating cell nuclear antigen (POL30) in *Saccharomyces cerevisiae*. Genetics. 2009;183:63-78. DOI:10.1534/genetics.109.103937.

[66] Zheng L, Zhou M, Chai Q, Parrish J, Xue D, Patrick SM, Turchi JJ, Yannone SM, Chen D, Shen B. Novel function of the flap endonuclease 1 complex in processing stalled DNA replication forks. EMBO Rep. 2005;6:83-89. DOI:10.1038/sj.embor.7400313.

[67] Lieber MR. The FEN-1 family of structure-specific nucleases in eukaryotic DNA replication, recombination and repair. BioEssays. 1997;19:233-240. DOI:10.1002/bies.950190309.

[68] Shen B, Nolan JP, Sklar LA, Park MS. Essential amino acids for substrate binding and catalysis of human flap endonuclease. J Biol Chem. 1996; 271: 9173-9176. DOI: 10.1074/jbc.271.16.9173.

[69] Moldovan GL, Pfander B, Jentsch S. PCNA, the Maestro of the Replication Fork. Cell. 2007;129:665-679. DOI: 10.1016/j.cell.2007.05.003.

[70] Strzalka W, Ziemienowicz A. Proliferating cell nuclear antigen (PCNA): A key factor in DNA replication and cell cycle regulation. Annals of Botany. 2011;107:1127-1140. DOI:10.1093/aob/mcq243.

[71] Boehm EM, Gildenberg MS, Washington MT. The Many Roles of PCNA in Eukaryotic DNA Replication. Enzymes. 2016;39:231-254. DOI:10.1016/bs.enz.2016.03.003.

[72] Anderson HJ, Vonarx EJ, Pastushok L, Nakagawa M, Katafuchi A, Gruz P, Di Rubbo A, Grice DM, Osmond MJ, Sakamoto AN, Nohmi T, Xiao W, Kunz BA. Arabidopsis thaliana Y-family DNA polymerase η catalyses translesion synthesis and interacts functionally with PCNA2. Plant J. 2008;55:895-908. DOI:10.1111/j.1365-313X.2008.03562.x.

[73] Markley NA, Young D, Laquel P, Castroviejo M, Moloney MM. Molecular genetic and biochemical analysis of *Brassica napus* proliferating cell nuclear antigen functionMolecular genetic and biochemical analysis of *Brassica napus* proliferating cell nuclear antigen function. Plant Mol Biol. 1997;34:693-700. DOI:10.1023/A:1005817220344.

[74] Kelman Z. PCNA: Structure, functions and interactions. Oncogene. 1997;14:629-40. DOI: 10.1038/sj.onc.1200886.

[75] De March M, Merino N, Barrera-Vilarmau S, Crehuet R, Onesti S, Blanco FJ, De Biasio A. Structural basis of human PCNA sliding on DNA. Nat Commun. 2017;8:13935. DOI: 10.1038/ncomms13935.

[76] Gomes X V., Burgers PMJ. Two modes of FEN1 binding to PCNA regulated by DNA. EMBO J. 2000;19:3811-3821. DOI:10.1093/emboj/19.14.3811.

[77] Sakurai S, Kitano K, Yamaguchi H, Hamada K, Okada K, Fukuda K, Uchida M, Ohtsuka E, Morioka H, Hakoshima T. Structural basis for recruitment of human flap endonuclease 1 to PCNA. EMBO J. 2005;24:683-693. DOI: 10.1038/sj.emboj.7600519.

[78] Singh SK, Roy S, Choudhury SR, Sengupta DN. DNA repair and recombination in higher plants: Insights from comparative genomics of arabidopsis and rice. BMC Genomics. 2010;11:443. DOI:10.1186/1471-2164-11-443.

[79] Lynch M, Conery JS. The evolutionary fate and consequences of duplicate genes. Science. 2000;290: 1151-1155. DOI: 10.1126/science.290.5494.1151.

[80] Lynch M, O'Hely M, Walsh B, Force A. The probability of preservation of a newly arisen gene duplicate. Genetics. 2001;159:1789-1804.

[81] Cheng F, Wu J, Cai X, Liang J, Freeling M, Wang X. Gene retention, fractionation and subgenome differences in polyploid plants. Nature Plants. 2018;4:258-268. DOI:10.1038/s41477-018-0136-7.

[82] Panchy N, Lehti-Shiu M, Shiu SH. Evolution of gene duplication in plants. Plant Physiol. 2016;171:2294-2316. DOI:10.1104/pp.16.00523.

[83] Li Z, Defoort J, Tasdighian S, Maere S, Van De Peer Y, De Smet R. Gene duplicability of core genes is highly consistent across all angiosperms. Plant Cell. 2016;28:326-344. DOI: 10.1105/tpc.15.00877.

Chapter 3

Sugarcane Breeding for Enhanced Fiber and Its Impacts on Industrial Processes

Pietro Sica

Abstract

For centuries, sugar has been virtually the only commercialized product derived from sugarcane. Traditionally, sugarcane breeding programs focused exclusively on the increase of the sucrose content, abandoning characteristics such as biomass yield and fiber content. Recently, sugarcane gained prominence also for its potential in terms of biomass production. As a result, some sugarcane breeding programs began to look for ways to increase fiber content and biomass yield instead of sugar content. In the 1980s, Alexander created the concept of energy cane. Here we review the changes in the sugarcane breeding programs related to enhanced fiber instead of sugar content. Compare the energy generation of energy cane with other biomass crops. Also, the recent changes in the biomass and biofuels scenario, focusing on topics as 2G ethanol and the RenovaBio program, from the Brazilian Government, which will give carbon credits to biofuels. Although several studies demonstrate its potential for biomass production, energy cane is still a new technology on an experimental scale and has been struggling to reach and establish on a commercial scale. However, policies and new technologies are increasing the demand for lignocellulosic material. Therefore, this chapter connects these points and shows the potential of this new plant material for the coming years.

Keywords: energy cane, bioenergy, 2G ethanol, RenovaBio, cane breeding, biomass

1. Introduction

Sugarcane is the most produced crop in the world, yielding about 1,890 million tons on approximately 26.8 million hectares. About 40.6% of the world's production was in Brazil. Asia (38.2%) also stands out, especially with countries such as India (18.4%) and China (6.5%), which are the second and third largest producers in the world, respectively. However, due to its high yields (70.6 tons/ha), the sugarcane harvested area is lower than other production main crops such as wheat, maize, and rice [1].

Dry matter of sugarcane is composed of sugar, mostly sucrose, and fiber (cellulose, hemicelluloses, and lignin) [2]. Studies have shown that sugarcane commercial hybrids stem is composed of about 14–18% sucrose and 12–15% fiber. The rest consists of water, minerals, and other substances [3, 4].

Worldwide, sugarcane is primarily grown as a source of sugar, providing around 70% of the world's sugar demand [5]. In 2018/2019, the world sugar production is expected to yield about 188 million metric tons of raw sugar, of which 68 million will be produced in Brazil and India [6]. In 2017, world ethanol production was about 27.05 billion gallons—the United States was the world's largest producer (58%), followed by Brazil (26%). However, the vast majority of US ethanol is produced from corn, while Brazil primarily uses sugarcane [7]. For this reason, traditionally, sugarcane breeding programs have focused on increasing the sugar content.

Sugarcane bagasse and straw have a high content of fiber and can be used for energy purposes according to four platforms: cogeneration, production of second-generation bioethanol, gasification to produce syngas, or generation of biogas, and pyrolysis to produce bio-oil and biochar [8–10]. In 2018, in the Brazilian final energy consumption sugarcane products represented 17.2% of all energy consumed in Brazil, of which 10.8% was from the sugarcane bagasse and 6.4% from ethanol, those products together were higher than mineral coal (14.4%), natural gas (11.4) and firewood (9.1%) [11].

Thus, sugarcane has great potential as a source of bioenergy. This review will present and discuss the traditional breeding programs that aim to increase sucrose content and the shift of this paradigm, focusing on the increase of biomass and fiber for bioenergy generation. In addition to the prospects for the use of biomass as a renewable and sustainable source of energy in the coming years.

2. Sugarcane breeding

Sugarcane is a crop that belongs to the genus *Saccharum L.* in the Poaceae family. Its genus includes six different species with variable sizes and numbers of chromosomes: *S. officinarum*, *S. spontaneum*, *S. robustum*, *S. barberi*, *S. sinense*, and *S. edule*. There are four genera closer to *Saccharum L.* that can readily interbreed (*Sclerostachya, Miscanthus, Narenga,* and *Erianthus*), forming the '*Saccharum* complex`. Three gene pools for sugarcane were proposed (**Table 1**).

One of the biggest challenges for taxonomists and molecular biologists is that '*Saccharum* complex' genera have a high level of polyploidy and aneuploidy, that is, an unbalanced number of chromosomes. Therefore, the complexity and size of the sugarcane genome are limitations in genetic improvement [12, 14]. Excessive non-flowering is one of the desirable characteristics of a sugarcane cultivar because flowering causes a pithing process in the stalks. However, flowering is a crucial characteristic in a breeding program, making it necessary for breeding stations to be built in specific locations where these phenotypes may flower regularly and have fertility. Another challenge for sugarcane breeders is the time. In Brazil, the breeding program takes from 11 to 13 years starting from the first crossbreeding performed until the release of a cultivar, since it is necessary to evaluate the clones on diseases and pests, as well as their productivity in different environments [15].

Gene pool	Genera examples	Ease of crossing	Hybrids
GP-1	*S. officinarum* clones, *S. robustum*, *S. spontaneum*, *Erianthus* and *Miscanthus*	Easy	Fertile
GP-2	Remainder of '*Saccharum* complex'	Some biological barriers make it more difficult	Tend to be sterile
GP-3	*Sorghum* and *Zea*	Needs techniques to enable gene transfer	Weak, lethal, or completely sterile

Table 1.
Proposed gene-pools for sugarcane and potential genera to be used in sugarcane breeding programs (adapted from [12, 13]).

2.1 Traditional sugarcane breeding

Until the 19th century, the most cultivated species was *Saccharum officinarum* because of its high sugar content [12]. Thereafter, sugarcane breeders that were interested in increasing disease resistance and yield crossed *S. officinarum* with a wild and vigorous relative, *S. spontaneum*, and then backcrossed the hybrids to *S. officinarum* [16]. *The* hybridization of both genera resulted in modern cultivars with chromosome numbers ranging from 100 to 130 [17], of which 80% originate from *S. officinarum* and 10–15% from *S. spontaneum*, with about 5–10% being recombinant chromosomes [18, 19].

S. officinarum was first found growing in gardens in the aborigines of New Guinea, a humid and high-temperature region. Later, it began to be used as raw material for sugar production, playing an important social, economic, and cultural role during the colonial period [20]. *S. spontaneum* is a grassy wild species found in diverse environments from Africa to Southeast Asia and the Pacific Islands. Since it evolved in such different environments, it has a wide gene pool, can adapt to different climate characteristics and is resistant to diseases to which *S. officinarum* is susceptible [21, 22].

Later, during the 20th century, sugarcane breeding programs were expanded, and all the efforts were still concentrated on increasing the sugarcane yield for sugar. Current levels of sugar yields are difficult to overcome, especially when considering the management systems and the carbon partition between the accumulation of sucrose and plant growth [23, 24]. The initial success of some breeding programs and subsequent stagnation in the genetic gains of sugarcane can be seen in the yield variation in the US, Brazil, and the World in the last fifty years, especially in the last decade [25].

In 1989, [22] described the 'modern' (1890–1989) sugarcane breeding process by dividing it into three phases. The first phase involved crossing and selecting among *S. officinarum* clones. By that time several clones were used by sugar industries worldwide. Those clones had commercial milling qualities such as sugar content, low fiber, and low impurity levels. However, they were susceptible to some diseases and had low vigor and ratooning performance. The second phase required developing interspecific hybrids by crossing the selected clone in the first phase with other species, which is normally part of the '*Saccharum* complex'. *S. spontaneum* has high adaptability to diverse environments, disease resistance, high vigor, and ratooning capacity. Because of this, it was mostly used in the interspecific cross with *S. officinarum*. To increase sugar content and stalk size, breeders used a process called "nobilization". Nobilization was based on backcrossing the initial hybrids with *S. officinarum* clones to increase the sugar content and the stalk size. The third phase was the multiplication and exploitation of the hybrids obtained in the second phase [12].

This breeding process, however, led to the narrowing of the genetic base of sugarcane breeding programs [21, 22]. Backcrossing sugarcane hybrids with *S. officinarum,* despite increasing the sugar content reduces fiber content and vigor, as the varieties are more susceptible to abiotic stresses and diseases [26]. Thus, the commercial average yield of sugarcane is about 25% of the potential field yield of fresh biomass in optimum conditions, 400 tons per hectare [27]. Because of these concerns, the interest in genetic diversity increased and sugarcane breeders saw a potential opportunity to introgress new genes into commercial hybrids [28]. The '*Saccharum* complex' fiber and sugar content is presented in **Table 2** [29].

2.2 Sugarcane breeding for fiber and the energy cane concept

After more than 100 years of looking to increase sugar yield, by the beginning of this century sugarcane breeding programs started to search for a new type of cane,

Species (n)	Sucrose (%)	Reducing sugar (%)	Fiber (%)
Erianthus maximus (3)	2.24 ± 0.44	0.73 ± 0.23	26.4 ± 0.9
Erianthus arundinaceus (2)	0.62 ± 0.16	0.61 ± 0.17	30.3 ± 0.3
Miscanthus floridulus (5)	3.03 ± 0.56	0.79 ± 0.24	51.0 ± 2.0
Saccharum spontaneum (30)	5.35 ± 0.38	1.66 ± 0.06	31.8 ± 0.9
Saccharum robustum (10)	7.73 ± 0.83	0.27 ± 0.02	24.8 ± 1.6
Saccharum sinense (2)	13.45 ± 0.02	0.38 ± 0.08	12.8 ± 2.0
Saccharum officinarum (25)	17.48 ± 0.35	0.32 ± 0.02	9.8 ± 0.4

n = number of evaluated accessions.

Table 2.
Levels of sucrose, reducing sugars, and fiber in access of ancestral genera and species of sugarcane [9, 29].

focusing on high yield and fiber for bioenergy generation [9, 30]. Now the goal of some breeding programs is to produce hardy plants with less juice and higher fiber productivity. A new introgression process is being used, this time replacing sugar plants with fibrous plants (see **Tables 1** and **2**). Reducing the sugar and increasing the fiber content would make plants more rustic, bringing about economic and environmental benefits as well as increased resistance to pests and diseases [9].

In the 1970s, in Puerto Rico, Alexander [31] already drew attention to "changing the focus on the 'qualitative side' features of sugarcane, as sugar yield, to the 'quantitative side,' such as green yield." He was one of the first breeders to use the energy cane concept in 1985. By that time, Alexander affirmed that if the energy cane were harvested with leaves and the top, it would increase the total biomass by 100%, with a penalty of 25–35% of sucrose [30, 31]. Energy cane can also be cultivated on marginal soil, optimizing the use of land [30] because it has a deeper root system, exploring better the soil's nutrients and water. For this reason, it has a big potential to be cultivated in 32 million hectares of degraded pastures in Brazil, more than the agricultural area of Europe [32].

Energy cane also differs from sugarcane in terms of the stalk morphology and the population. **Figure 1** summarizes the differences between energy cane and sugarcane on the stalk morphology and population. Energy cane stalk height is from 4 to 6 m, whereas sugarcane stalk is smaller, from 2 to 2.5 m [33, 34]. However, the sugarcane stalk is thicker than the energy cane. The diameter of the sugarcane stalk is, on average, 3.5 cm, whereas the energy cane ranges from 1.5 to 2 cm [32]. In terms of population, the cultivation of energy cane can have from two to three times more plants per area when compared to the traditional cultivation systems of sugarcane [32].

According to [35] to obtain energy cane cultivars, breeding programs should:

i. maintain a germplasm collection with high genetic diversity in sugar and fiber content.

ii. perform genetic crosses between modern hybrids of sugarcane (high sugar) and accesses of the genus Saccharum (high fiber).

iii. produce large amounts of seedlings and select superior individuals.

To better understand sugarcane as an energy crop and to facilitate well-focused and effective genetic improvement programs, [36] classified the energy cane into three distinctive types. The first is sugarcane, composed primarily of sugars, with

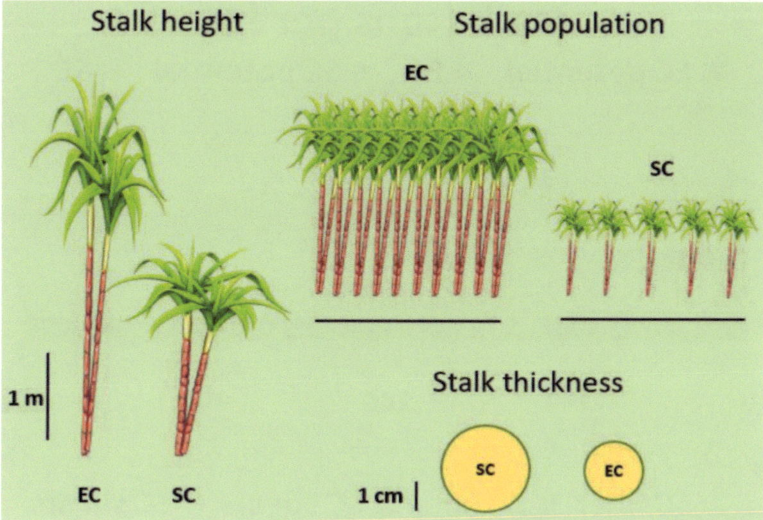

Figure 1.
Comparison between the characteristics of the stalks of sugar cane and energy cane: height, thickness, and population. EC: energy cane and SC: sugarcane. Sources: adapted from [32–34].

a juice with a high concentration of sucrose, high purity, which can be used in both sugar and ethanol production. Type I energy cane is like conventional sugarcane but has higher fiber content and less sucrose. Its juice's purity is lower and is not recommended for sugar production. Type II energy cane has only a marginally content of sugar and higher fiber than that of Type I, and it should be used exclusively for biomass production.

One of the pioneer programs for introgression in energy cane started in the 1980s in Barbados. One hybrid, WI79460, achieved yields of 112 tons of biomass and 46 tons of dry mass per hectare, a gain of 73% when compared to a commercial cultivar used in sugar production, B77602 (26.7 tons of dry mass per hectare). WI79460 also had a relatively high sucrose production per hectare of 10.4 tons, 77.03% of the sucrose produced by the B77602 [4, 9].

In Brazil, CanaVialis has started an introgression program for fiber content, and preliminary studies already show enormous potential. A commercial hybrid was crossed with *S. spontaneum* and the best F_1 clone was reproduced and compared with one of the most used commercial hybrids in Brazil, RB72454. The number of stalks per linear meter was 40, which is a high value when compared with the commercial hybrid that has 14 stalks per linear meter. The total fiber per hectare was also higher. The selected clone produced 40.25 tons of fiber per hectare, which was 136% higher than that of the commercial hybrid (17 tons per hectare). However, the sugar production per hectare and the purity of the extracted juice was lower in the energy cane [3]. **Figure 2** compares the production and potential production of energy cane and sugarcane in tons of stalks, fiber, and sugars per hectare.

Since 2001, USDA scientists at the Sugarcane Research Laboratory in Houma, Lousiana, are assessing the energy potential of high-fiber sugarcanes. In 2007, in the Louisiana sugar belt, three sugarcane varieties with fiber content higher than 16% were released for use as feedstocks for the production of bioenergy. These varieties, however, are disqualified for use in commercial sugar production operations [37].

Based on the above, energy cane is a result of recent breeding programs and may be an alternative to be integrated with the traditional ethanol and sugar production processes.

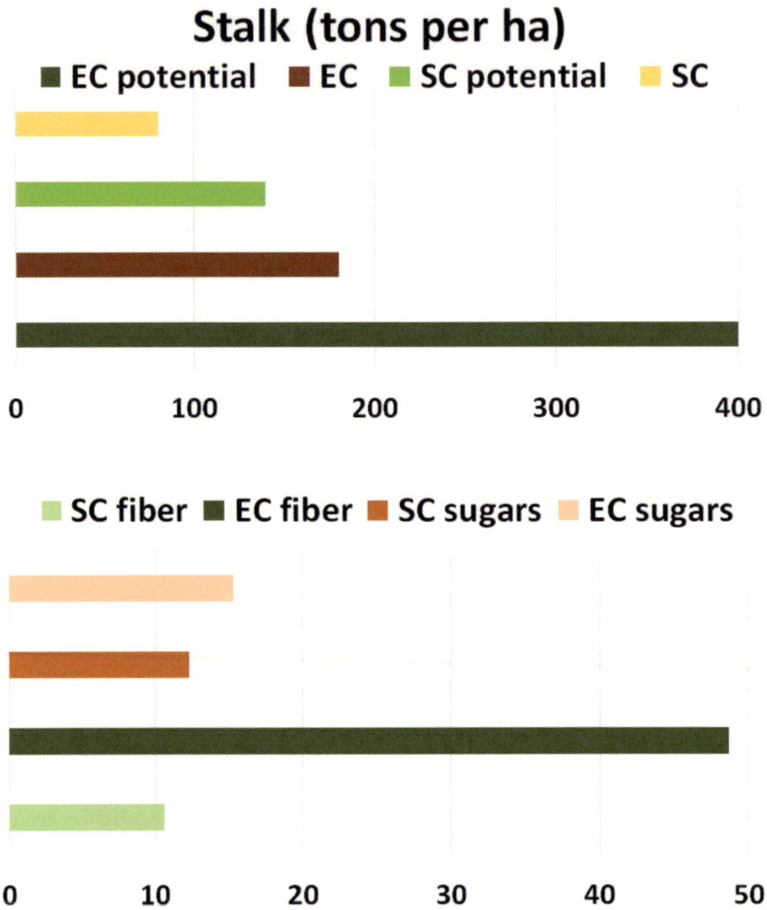

Figure 2.
Energy cane and sugarcane biomass, fiber, and sugar production and potential production per hectare. EC: energy cane and SC: sugarcane. Sources: adapted from [4, 25, 30, 32].

3. Fiber as a source of energy

To be considered a sustainable biomass source, the plant should meet humanity's energy needs, without competing with food production. For that, fibrous plants should be prioritized, instead of starch and oilseed plants [38]. In 2014, in the review "The potential of energy cane as the main biomass crop for the cellulosic industry", [9] listed five characteristics of the use of fibrous plants as biomass, citing [38–43]:

i. C4 plants with high efficiency in assimilate the solar energy and convert it into biomass, with the less possible amount of water, nutrients, and other inputs.

ii. possibility of application of agricultural technology in large-scale production.

iii. perennial but growth and long-term canopy to allow harvest during most of the year.

iv. easily and efficiently processed into usable forms of energy.

v. sustainable economically and environmentally, that is, not compete with food production, being able to be cultivated in marginal lands, with social-benefit consequences and having a high rate of carbon (C) balance.

The fibrous stem with low sugar content makes the energy cane more valuable to produce 2G ethanol and bioelectricity. However, the sugarcane price in Brazil is determined based on the sugar content and purity, and the sugarcane's end use, either sugar or ethanol [44, 45]. Thus, this payment method still encourages the production of the traditional sugarcane, to produce sugar and ethanol, being an obstacle to the cultivation of energy cane by sugarcane suppliers. Therefore, it is still necessary to have policies and to develop a payment method that considers the fiber content and the production of bioelectricity and 2G ethanol.

3.1 Energy cane as biomass and 2G ethanol

In the last decades, due to the increasing prices of electricity sold to the grid, specific incentive policies, and public-private initiatives, the Brazilian sugarcane industry increased its focus on cogeneration [35]. From 2013, the amount of commercialized surplus electricity from the plant is higher than the self-consumed, achieving a ratio of 60% and 40% in 2015 [46].

Cellulosic ethanol is a biofuel produced from extracted cellulose from the fibers of a vegetable. In the case of sugarcane, the primary input used in Brazil to produce 2G ethanol is obtained by processing the bagasse after extracting the juice or even the straw. As about half of the energy from sugarcane is present in its lignocellulosic fibers (bagasse and straw), it would be possible to produce more ethanol and electricity with the same amount of material and planted area [47, 48]. The energy cane can provide more fiber for the industrial process, supplying the cogeneration and bioelectricity production, and increasing the amount of bagasse that can be used for the 2G ethanol production. In Brazil, three companies already started to use the second-generation ethanol technology, with a total capacity of 124 million liters of second-generation ethanol per year:

i. Raízen, in a plant located in Piracicaba-SP, since 2014 producing ethanol from bagasse and straw, being able to increase by 50% the ethanol production without expanding the area of cultivation, producing biofuel even during the off-season for sugarcane, from December until March, and reducing carbon emissions during production, creating a cleaner fuel. The plant can produce 42 million liters of ethanol [49];

ii. GramBio, also started the commercial plant in 2014. The plant, Bioflex 1, is in Sao Miguel dos Campos, in the state of Alagoas, in the Northeast region, and its initial production capacity is 82 million liters of ethanol per year from energy cane bagasse and straw [50].

iii. CTC (Sugarcane Technology Center) has a pilot plant in Sao Manoel-SP, producing about 3 million liters of 2G ethanol per year from sugarcane straw and bagasse [51].

The 2G ethanol substrate is more diluted than the 1G ethanol, generating higher volumes of vinasse per liter of ethanol produced with a chemical composition different than the traditional 1G vinasse [52–54]. However, due to the high acetic acid concentration and to the low content of furans, which can inhibit the anaerobic digestion process, biogas production from 2G ethanol vinasse can have satisfactory performance, when compared to 1G ethanol vinasse [54]. The filter cake is rich in nutrients and can also be used integrated with the vinasse to increase the biogas production efficiency [55]. If ensiled, the energy cane can be stored for up to six months and be used to feed the biogas process in the period when the ethanol plant is not in operation [56]. Also, according to [56], in terms of energy generation per cultivated area, the production of biogas from sugarcane bagasse is more advantageous than the production of 2G ethanol Thus, it is also possible to consider a biogas production plant integrated with a 2G ethanol plant to increase the bioenergy generation.

The second-generation ethanol has great potential to increase the biofuel production in Brazil without increasing the sugarcane cultivation area [57]. However, this technology is not considered viable yet, needing to overcome some challenges, including the pretreatment and hydrolysis conditions to release the fermentable sugars [58]. Although these challenges need to be faced, the second-generation ethanol from sugarcane and energy cane can become a reality in Brazil for three reasons:

i. the Federal Government has incentive programs, and the commercialization of carbon credits is going to be one more economical and environmental advantage for this product [59–61];

ii. this technology is still in a learning curve period, and it is expected that 2G technologies are going to become more competitive in the future [62, 63];

iii. the cultivation of energy cane will boost the availability of fiber in the industry [64].

However, to process the energy cane in the ethanol plant is still challenging. Different conditions for milling and pretreatments may be required [9, 37]. Although the primary energy and sugar production per hectare is higher for energy cane when compared to the traditional sugarcane, the sugar concentration in the juice is lower and fiber content is higher. All the extraction process in the plant are adapted for sugarcane, thus, these differences raise concerns about processing energy cane, as mentioned by [65]:

i. sugar extraction: still need to be studied which process is better for it: mill or diffuser; the amount of water required for imbibition; the energy requirement for the extraction; and the extraction efficiency, which is expected to be reduced as the % of fiber increases;

ii. steam consumption; more imbibition water and increase in % fiber will require more steam for the processing of energy cane. However, it is also expected that a greater amount of fiber will be supplied to the cogeneration, having a positive balance in the steam generation:consumption;

iii. processing: needs to be reevaluated to maximize the products revenue

3.2 RenovaBio and energy cane

It is expected that the second-generation ethanol production in Brazil will increase and achieve about 2 billion liters in 2030, almost 20 times more than the current production [59] due to a new Federal Government program, the RenovaBio. In the RenovaBio, biofuel producers will receive one financial title

	Sugarcane[a]	Energy cane[a]	Sweet sorghum[b]	Eucalyptus[c]
Crop cycles (months)	10 to 12	10 to 15	3.5	96
Number of cycles year^{-1}	1	1	2	0.125
Yield (t ha^{-1} year^{-1})	70	200	60	24
Brix (% juice)	13 to 15	10 to 12	11 to 13	—
Fiber (% cane)	13.5	26.7	13	—
Biomass (t ha^{-1} year^{-1})	17.5	50	15.6	24
Calorific power (Mcal t^{-1})	2,275	2,275	3,281	4,600
Mcal ha^{-1} year^{-1}	39,813	113,750	51,184	110,400

[a] *[3, 25, 66–68].*
[b] *[66, 69].*
[c] *[70–72].*

Table 3.
Comparison of the potential energy generation per hectare per year among biomass crops.

equivalent to carbon credits called CBIO, which corresponds to one ton of CO_2 that is no longer emitted due to the biofuel production. Fuels distributors will have an obligation to buy CBIOs and it will also be available to any interested investor [60, 61]. The energy cane juice contains 9.8% of fermentable sugars, less than half of the traditional sugarcane. With these incentives of the Brazilian government and advances on the 2G technologies, the energy cane can potentially triple the productivity of biomass per hectare and reduce the production costs, also increasing the production of biofuel in the same area. The theoretical ethanol yields of sugarcane is 3,609 kg per hectare, while the energy cane are 12,938 kg of ethanol per hectare [61, 66]. However, it is important to find ways to ensure that these incentives reach the sugarcane producers, and not just stop in the sugarcane industry.

Regarding the amount of energy per area per year, the energy cane has almost four times more energy content than the traditional sugarcane, more than double of the sweet sorghum content, and almost the same as eucalyptus. However, the energy cane also provides the juice with a considerable amount of fermentable sugars and the eucalyptus cycle can take more than eight years (**Table 3**). The energy cane also has higher yields than the elephant grass and erianthus [73, 74].

4. Final remarks

Sugarcane breeding programs began in the 19th century when a focus on increasing sugar yield surfaced and have continued until recently. However, in the last few decades, various policies around the world have started aiming to reduce dependence on petroleum and other fossil energy sources. Thus, many breeders around the world have turned their attention to increasing productivity and fiber content. In this scenario, Brazil appears as the ideal country to start its commercial cultivation, because, in addition to being tropical, the RenovaBio program will give financial incentives to increase the production of second-generation ethanol. The energy cane is a crop with higher fiber production potential in marginal soil. In addition to that, the literature presented in this chapter shows that the energy cane can be stored and used to keep a continuous biogas production from December until March when the ethanol plant is not operating, and vinasse is not being produced.

However, energy cane faces several challenges to be implemented commercially. In the RenovaBio context, it is still necessary to find ways that the incentives do not end in the industry. CBIOs should also achieve producers, as a form of incentive for more sustainable production, closing the producer-industry-consumer cycle, and stimulating a more sustainable biomass production in the field and, consequently, the adoption of energy cane. The industrial processes and sugarcane management in the field have already been established for decades, and the implementation of energy cane would imply several changes throughout the chain—this is still a reason for resistance from producers and mill managers. Also, the low purity of its juice does not allow it to be used for sugar production and the current sugarcane method valorizes high sugar content and purity, to produce sugar and ethanol. In this sense it is also necessary to develop a new payment method, considering the fiber content in order to stimulate producers to adopt the energy cane. In addition, cellulosic ethanol is still a very new technology and needs adjustments to reach industrial scales and become profitable.

Energy cane is still an experimental technology and its cultivation is starting to be adopted on larger scales; however, it demonstrates the potential to be expanded to commercial scales. To do this, further steps need to be taken in breeding programs and new technologies, and changes will be necessary for its implementation in the field as well as its processing in the industry.

Author details

Pietro Sica
University of Sao Paulo, Piracicaba, Sao Paulo, Brazil

*Address all correspondence to: pietros0394@gmail.com

IntechOpen

© 2021 The Author(s). Licensee IntechOpen. This chapter is distributed under the terms of the Creative Commons Attribution License (http://creativecommons.org/licenses/by/3.0), which permits unrestricted use, distribution, and reproduction in any medium, provided the original work is properly cited.

References

[1] FAOSTAT. Beans, dry [Internet]. 2017 [cited 2020 Jan 7]. Available from: http://www.fao.org/faostat/en/#data/QC

[2] Sreenivasan T V., Nair N V. Catalogue on Sugarcane Genetic Resources III. Saccharum officinarum L. 1st ed. Coimbatore, India: ICAR; 1991.

[3] Matsuoka S, Bressiani J, Maccheroni W, Fouto I. Sugarcane bioenergy. In: Sugarcane: Agricultural Production, Bioenergy and Ethanol. Elsevier Inc.; 2015. p. 383-405.

[4] Rao PS, Davis H, Simpson C. New sugarcane varieties and year round sugar and ethanol production with bagasse-based cogeneration in Barbados and Guyana. XXVI Congr Int Soc Sugar Cane Technol ICC, Durban, South Africa, 29 July - 2 August, 2007. 2007;1169-76.

[5] Hofer R. Sugar- and Starch-Based Biorefineries. Ind Biorefineries White Biotechnol Elsevier, Amsterdam. 2015;(Chapter 4A):157-235.

[6] USDA. Sugar: Wold Markets and Trades [Internet]. 2018 [cited 2019 Nov 12]. Available from: http://www.fas.usda.gov/commodities/sugar

[7] RFA. Annual Ethanol Production | Renewable Fuels Association [Internet]. 2019 [cited 2019 Nov 12]. Available from: https://ethanolrfa.org/statistics/annual-ethanol-production/

[8] Buckeridge MS, De Souza AP, Tavares EQP, Cambler AB. Sugarcane cell wall structure and degradation: From monosaccharide analyses to the glycomic code. Advances of Basic Science for Second Generation Bioethanol from Sugarcane Springer International Publishing; Jan 1, 2017 p. 7-19.

[9] Carvalho-Netto O V., Bressiani JA, Soriano HL, Fiori CS, Santos JM, Barbosa GV, et al. The potential of the energy cane as the main biomass crop for the cellulosic industry [Internet]. Springer International Publishing; Dec 21, 2014 p. 20. Available from: http://www.chembioagro.com/content/1/1/20

[10] Dias MOS, Junqueira TL, Cavalett O, Cunha MP, Jesus CDF, Rossell CEV, et al. Integrated versus stand-alone second generation ethanol production from sugarcane bagasse and trash. Bioresour Technol. 2012 Jan;103(1):152-161.

[11] MME. National Energy Balance, based on 2018 [Internet]. Ministry of Mines and Energy. 2019 [cited 2020 Feb 6]. p. 69. Available from: http://www.epe.gov.br/sites-pt/publicacoes-dados-abertos/publicacoes/PublicacoesArquivos/publicacao-377/topico-470/Relatório Síntese BEN 2019 Ano Base 2018.pdf

[12] Daniels J, Roach B. Taxonomy and Evolution. In: Heinz GD, editor. Sugarcane Improvement Through Breeding. 1st ed. Amsterdam: Elsevier; 1987. p. 7-84.

[13] Harlan JR, deWet JMJ. On Ö. Winge and a Prayer: The origins of polyploidy. Bot Rev [Internet]. 1975 Oct [cited 2020 Dec 26];41(4):361-90. Available from: https://link.springer.com/article/10.1007/BF02860830

[14] Mariotti JA. Selection for sugar cane yield and quality components in subtropical climates. Int sugar J. 2001;103(1233):418-423.

[15] Barbosa MHP, da Silveira LCI. Breeding Program and Cultivar Recommendations. In: Santos F, Borem A, Caldas C, editors. Sugarcane: Agricultural Production, Bioenergy and Ethanol. 1st ed.

Amsterdam: Elsevier Inc.; 2015. p. 241-255.

[16] Berding N, Roach BT. Germplasm collection, maintenance, and use. In: Heinz GD, editor. Sugarcane Improvement Through Breeding. 1st ed. Amsterdam: Elsevier; 1987. p. 143-210.

[17] D'Hont A, Ison D, Alix K, Roux C, Glaszmann JC. Determination of basic chromosome numbers in the genus Saccharum by physical mapping of ribosomal RNA genes . Genome [Internet]. 1998 Apr 1 [cited 2020 Dec 26];41(2):221-5. Available from: https://cdnsciencepub.com/doi/abs/10.1139/g98-023

[18] D'Hont A, Grivet L, Feldmann P, Rao S, Berding N, Glaszmann JC. Characterisation of the double genome structure of modern sugarcane cultivars (Saccharum spp.) by molecular cytogenetics. Mol Gen Genet [Internet]. 1996 [cited 2020 Dec 26];250(4):405-13. Available from: https://link.springer.com/article/10.1007/BF02174028

[19] Piperidis G, Piperidis N, D'Hont A. Molecular cytogenetic investigation of chromosome composition and transmission in sugarcane. Mol Genet Genomics [Internet]. 2010 Jul 8 [cited 2020 Dec 26];284(1):65-73. Available from: http://www.bses.org.au

[20] Stevenson GC. Genetics and breeding of sugar cane. Genetics and breeding of sugar cane. London: Longmans, Green & Co. Ltd.; 1965. 284 p.

[21] Roach BT. A review of the origin and improvement of sugarcane. In: Copersucar International Sugarcane Breeding Workshop. Piracicaba, Brazil: Copersucar Technology Center; 1987.

[22] Roach BT. Origin and improvement of the genetic base of sugarcane. In: Australian Society of Sugar Cane Technologists. Palm Cove, Australia: ASSCT; 1989.

[23] Jackson PA. Breeding for improved sugar content in sugarcane. In: Field Crops Research. Elsevier; 2005. p. 277-290.

[24] Moore PH. Integration of sucrose accumulation processes across hierarchical scales: Towards developing an understanding of the gene-to-crop continuum. In: Field Crops Research. Elsevier; 2005. p. 119-135.

[25] FAO. FAOSTAT [Internet]. 2019 [cited 2019 Nov 10]. Available from: http://www.fao.org/faostat/en/#data/QC

[26] Waclawovsky AJ, Sato PM, Lembke CG, Moore PH, Souza GM. Sugarcane for bioenergy production: An assessment of yield and regulation of sucrose content. Plant Biotechnol J [Internet]. 2010 Apr 1 [cited 2020 Dec 26];8(3):263-76. Available from: https://onlinelibrary.wiley.com/doi/full/10.1111/j.1467-7652.2009.00491.x

[27] Cooper J. Potential Productivity of Field Crops under Different Environments. In: Smith WH, Banta SJ, editors. Experimental Agriculture. Philippines International Rice Research Institute; 1983. p. 529.

[28] Miller JD, Tai PYP. Use of Plant Introductions in Sugarcane Cultivar Development. In: Shands HL, Weisner LE, editors. Use of Plant Introductions in Cultivar Development Part 2 [Internet]. Madison, Wisconsin: CSSA; 1992 [cited 2020 Dec 26]. p. 137-49. Available from: http://doi.wiley.com/10.2135/cssaspecpub20.c6

[29] Bull T, Glasziou K. The Evolutionary Significance of Sugar Accumulation in Saccharum. Aust J Biol Sci [Internet]. 1963 [cited 2020 Dec 26];16(4):742. Available from: https://www.publish.csiro.au/bi/bi9630737

[30] Matsuoka S, Kennedy AJ, Santos EGD dos, Tomazela AL, Rubio LCS. Energy Cane: Its Concept, Development, Characteristics, and Prospects. Adv Bot. 2014;2014:1-13.

[31] Alexander AG. The energy cane alternative. Elsevier Sci Ltd Amsterdam. 1985;

[32] GranBio. What is energy cane? [Internet]. 2015 [cited 2021 Jan 7]. Available from: http://www.granbio.com.br/en/conteudos/energy-cane/

[33] Pokhrel P, Rajan N, Jifon J, Rooney W, Jessup R, da Silva J, et al. Agronomic performance of the lignocellulosic feedstock crop energy cane in the Texas Rolling Plains. Agron J [Internet]. 2020 Sep [cited 2021 Jan 7];112(5):3816-31. Available from: https://onlinelibrary.wiley.com/doi/10.1002/agj2.20370

[34] Santos LR. Productivity of cane-energy cultivar by DSSAT / CANEGRO model (in Portuguese) [Internet]. [Rio Largo]: Federal University of Alagoas; 2018 [cited 2021 Jan 7]. Available from: http://www.repositorio.ufal.br/handle/riufal/3376

[35] Barbosa GV de S, Dos Santos JM, Diniz CA, Cursi DE, Hoffmann HP. Energy cane breeding. In: Santos F, Rabelo S, Matos M, Eichler P, editors. Sugarcane Biorefinery, Technology and Perspectives. 1st ed. Amsterdam: Elsevier; 2020. p. 103-116.

[36] Tew TL, Cobill RM. Genetic improvement of sugarcane (Saccharum spp.) as an energy crop. In: Vermerris W, editor. Genetic Improvement of Bioenergy Crops [Internet]. 1st ed. New York: Springer; 2008 [cited 2020 Dec 26]. p. 273-94. Available from: http://sugarcane-breeding.tn.nic.in/genresources.htm

[37] Tew T, Hale A, Pan Y, Richard Jr E. Energy Cane Breeding and Selection in Louisiana - A Progress Report. [Internet]. International Society of Sugar Cane Technologists. 2009 [cited 2020 Mar 25]. Available from: https://www.ars.usda.gov/research/publications/publication/?seqNo115=240040

[38] Sticklen MB. Plant genetic engineering for biofuel production: Towards affordable cellulosic ethanol. Nat Rev Genet [Internet]. 2008 Jun [cited 2020 Dec 26];9(6):433-43. Available from: www.nature.com/reviews/genetics

[39] Hill J, Nelson E, Tilman D, Polasky S, Tiffany D. Environmental, economic, and energetic costs and benefits of biodiesel and ethanol biofuels. Proc Natl Acad Sci U S A [Internet]. 2006 Jul 25 [cited 2020 Dec 26];103(30):11206-10. Available from: www.pnas.orgcgidoi10.1073pnas.0604600103

[40] Rubin EM. Genomics of cellulosic biofuels. Nature [Internet]. 2008 Aug 14 [cited 2020 Dec 26];454(7206):841-5. Available from: http://www.checkbiotech.org/green_News_Biofuels.aspx?infoId515100

[41] Sticklen MB. Feedstock crop genetic engineering for alcohol fuels. Crop Sci [Internet]. 2007 Nov 1 [cited 2020 Dec 26];47(6):2238-48. Available from: https://acsess.onlinelibrary.wiley.com/doi/full/10.2135/cropsci2007.04.0212

[42] Coombs J. Sugar-cane as an Energy Crop. Biotechnol Genet Eng Rev [Internet]. 1984 [cited 2020 Dec 26];1(1):311-46. Available from: https://www.tandfonline.com/action/journalInformation?journalCode=tbgr20

[43] Schmer MR, Vogel KP, Mitchell RB, Perrin RK. Net energy of cellulosic ethanol from switchgrass. Proc Natl Acad Sci U S A [Internet]. 2008 Jan

15 [cited 2020 Dec 26];105(2):464-9. Available from: www.pnas.orgcgidoi10.1073pnas.0704767105

[44] CONSECANA. Sugarcane Payment Manual - 5th ed (in Portuguese). Piracicaba; 2006.

[45] Martines-Filho J, Burnquist HL, Vian CEF. Bioenergy and the Rise of Sugarcane-Based Ethanol in Brazil. Choices. 2006 Jul 18;21(2):91-96.

[46] UNICA. CANE'S BIOELECTRICITY IN NUMBERS-JANUARY 2017 (in Portuguese) [Internet]. Sao Paulo; 2017 [cited 2020 Feb 9]. Available from: http://www.siamig.com.br/uploads/a309a426211622d6594e960fa9202c12.pdf

[47] BNDES. Bioethanol from Sugarcane: Energy for a Sustainable Development [Internet]. 1st ed. Nogueira L, Seabra J, Best G, Leal M, Poppe M, editors. Rio de Janeiro, BNDES/ .... Rio de Janeiro: Banco Nacional de Desenvolvimento Econômico e Social; 2008 [cited 2020 Feb 9]. 1-316 p. Available from: http://web.bndes.gov.br/bib/jspui/handle/1408/2002

[48] Lorenzi BR, Andrade THN de. Second Generation Ethanol in Brazil: Policies and Social-Technical Networks (in Portuguese). Rev Bras Ciências Sociais. 2019;34(100):1-19.

[49] Raízen. Second generation ethanol [Internet]. 2014 [cited 2020 Feb 9]. Available from: http://www.raizen.com.ar/en/energy-future/renewable-energy-technology/second-generation-ethanol

[50] Granbio. GranBio begins producing second-generation ethanol [Internet]. Sao Miguel dos Campos; 2014 [cited 2020 Feb 9]. Available from: http://www.granbio.com.br/en/wp-content/uploads/sites/2/2014/09/partida_english.pdf

[51] CTC. Cellulosic Ethanol [Internet]. Sugarcane Technology Center. [cited 2021 Jan 7]. Available from: https://ctc.com.br/en/cellulosic-ethanol/

[52] Jardine JG, Dispato I, Peres MR. Considerations on Lignocellulosic Bioethanol to Support the Content Development of the Agroenergy Knowledge Tree (in Portuguese) [Internet]. Vol. 95. Campinas: EMBRAPA; 2009 [cited 2020 Dec 26]. 1-32 p. Available from: www.cnptia.embrapa.br

[53] Monlau F, Sambusiti C, Barakat A, Quéméneur M, Trably E, Steyer JP, et al. Do furanic and phenolic compounds of lignocellulosic and algae biomass hydrolyzate inhibit anaerobic mixed cultures? A comprehensive review. Biotechnol Adv. 2014 Sep 1;32(5):934-951.

[54] Silverio MS, Calegari RP, Leite GMFL, Martins BC, Silva EA da, Piotrovski Neto J, et al. Biogas Production from Second Generation Ethanol Vinasse. In: Silva DAS, editor. Agronomist: Elo da Cadeia Produtiva. 1st ed. Belo Horizonte: Atena Editora; 2019. p. 174-184.

[55] Janke L, Leite A, Nikolausz M, Schmidt T, Liebetrau J, Nelles M, et al. Biogas Production from Sugarcane Waste: Assessment on Kinetic Challenges for Process Designing. Int J Mol Sci [Internet]. 2015 Aug 31 [cited 2020 Feb 9];16(9):20685-703. Available from: http://www.mdpi.com/1422-0067/16/9/20685

[56] Hoffstadt K, Pohen GD, Dicke MD, Paulsen S, Krafft S, Zang JW, et al. Challenges and Prospects of Biogas from Energy Cane as Supplement to Bioethanol Production. Agronomy [Internet]. 2020 Jun 9 [cited 2020 Dec 26];10(6):821. Available from: https://www.mdpi.com/2073-4395/10/6/821

[57] Goldemberg J, Mello FFC, Cerri CEP, Davies CA, Cerri CC.

Meeting the global demand for biofuels in 2021 through sustainable land use change policy. Energy Policy. 2014 Jun 1;69:14-18.

[58] Santos FA, De Queiróz JH, Colodette JL, Fernandes SA, Guimaraes VM, Rezende ST. Potential of sugarcane straw for ethanol production. Quim Nova. 2012;35(5):1004-1010.

[59] MME. Scenarios for Ethanol Supply and Otto Cycle Demand - 2018-2030 (in Portuguese) [Internet]. Rio de Janeiro; 2018 [cited 2020 Feb 9]. Available from: http://www.epe.gov.br/sites-pt/publicacoes-dados-abertos/publicacoes/PublicacoesArquivos/publicacao-255/topico-392/EPE-DPG-SGB-Bios-NT-01-2017-r0_Cenarios_de_Oferta_de_Etanol.pdf

[60] ANP - Brazilian National Agency of Petroleum, Natural Gas and B. RenovaCalc [Internet]. 2019 [cited 2019 Nov 10]. Available from: http://www.anp.gov.br/producao-de-biocombustiveis/renovabio/renovacalc

[61] Grassi MCB, Pereira GAG. Energy-cane and RenovaBio: Brazilian vectors to boost the development of Biofuels. Ind Crop Prod. 2019 Mar 1;129:201-205.

[62] Wang L, Quiceno R, Price C, Malpas R, Woods J. Economic and GHG emissions analyses for sugarcane ethanol in Brazil: Looking forward. Vol. 40, Renewable and Sustainable Energy Reviews. Elsevier Ltd; 2014. p. 571-582.

[63] BNDES. From promise to reality: how cellulosic ethanol can revolutionize the sugar cane industry - an assessment of the competitive potential and public policy suggestions (in Portuguese). 1st ed. Yabe Milanez A, Nyko D, Soares Valente M, Cunha De Sousa L, Charles AB, Farias De Jesus D, et al., editors. Vol. 41, BNDES Setorial. Rio de Janeiro: Banco Nacional de Desenvolvimento Econômico e Social; 2015. 237-294 p.

[64] Lorenzi BR, de Andrade THN. Second generation ethanol in Brazil: Policies and socio-technical networks. Rev Bras Ciencias Sociais [Internet]. 2019 [cited 2021 Jan 7];34(100). Available from: https://orcid.org/0000-0001-8486-0726

[65] Leal MRL V. The Potential of Sugarcane as an Energy Source. Proc Int Soc Sugar Cane Technol. 2007;26:23-34.

[66] Kim M, Day DF. Composition of sugar cane, energy cane, and sweet sorghum suitable for ethanol production at Louisiana sugar mills. J Ind Microbiol Biotechnol. 2011 Jul;38(7):803-807.

[67] Silva MB, Morais AS. Energy evaluation of sugarcane bagasse in different humidity levels and degrees of compaction (in Portuguese). ENEGEP, Rio Janeiro, Brazil. 2008;

[68] Teixeira SR, Arenales A, de Souza AE, Magalhães R da S, Peña AFV, Aquino D, et al. Sugarcane Bagasse: Applications for Energy Production and Ceramic Materials. J Solid Waste Technol Manag [Internet]. 2015 Aug 1;41(3):229-38. Available from: http://openurl.ingenta.com/content/xref?genre=article&issn=1088-1697&volume=41&issue=3&spage=229

[69] Jóvér J, Antal K, Zsembeli J, Blaskó L, Tamás J. Assessment of gross calorific value of crop and bio-energy residues. Res Agr Eng [Internet]. 2018 [cited 2020 Jun 17];64(3):121-7. Available from: https://doi.org/10.17221/13/2017-RAE

[70] Barotto AJ, Monteoliva S, Gyenge J, Martínez-Meier A, Moreno K, Tesón N, et al. Wood density and anatomy of three eucalyptus species: Implications for hydraulic conductivity. For Sys. 2017;26(1).

[71] Carneiro A de CO, Castro AFNM, Castro RVO, Santos RC dos, Ferreira LP, Damásio RAP, et al. Energetic potential of Eucalyptus sp. depending on age and different genetic materials (in Portuguese). Rev Árvore. 2014;38(2):375-381.

[72] Magaton AS, Colodette JL, Piló-Veloso D, Gomide JL. Behavior of Eucalyptus Wood Xylans across Kraft Cooking. J Wood Chem Technol [Internet]. 2011 Feb 28 [cited 2019 Nov 10];31(1):58-72. Available from: http://www.tandfonline.com/doi/abs/10.1080/02773813.2010.484123

[73] El Bassam N. Energy plant species: their use and impact on environment and development. London: James & James; 1998. 321 p.

[74] Woodard KR, Prine GM. Dry Matter Accumulation of Elephantgrass, Energycane, and Elephantmillet in a Subtropical Climate. Crop Sci [Internet]. 1993 Jul 1 [cited 2020 Dec 26];33(4):818-24. Available from: https://acsess.onlinelibrary.wiley.com/doi/full/10.2135/cropsci1993.0011183X003300040038x

Chapter 4

Physicochemical Properties of Sugarcane Industry Residues Aiming at Their Use in Energy Processes

Julia M. de O. Camargo, Jhuliana Marcela Gallego Ríos, Graziella C. Antonio and Juliana T.C. Leite

Abstract

According to the US Department of Agriculture, sugarcane global production for marketing year (MY) 2020/21 will forecast up 22 million tons in comparison with 2019/2020 MY, reaching 188 million tons (raw value), due to higher production in Brazil, India and Thailand. These countries alternate sugarcane uses for obtaining sugar, ethanol and other products, generating near to 152 million tons of residues. In a circular economy context, the reuse of the sugarcane industry by-products is desired. Nowadays, bagasse and, sometimes, straw are used for energy recovery through combustion, while filter cake and vinasse are commonly used for soil fertilization. However, while bagasse and straw present potential for energy recovery through the thermochemical route, vinasse and filter cake are better applied in anaerobic digestion processes to produce biogas and biofertilizer. These treatments, when correctly employed, can improve the performance of sugarcane industry by diversifying its energy sources and products. For this, the correct design of equipment and processes is essential, which requires the knowledge of physical and chemical properties of sugarcane industry's by-products. In this context, the chapter goal is to present an updated literature review for these properties, considering their use in energy recovery processes.

Keywords: waste-to-energy, agricultural wastes, physicochemical characterization, bagasse, sugarcane straw, vinasse, filter cake

1. Introduction

According to the United States Department of Agriculture, sugarcane global production for marketing year (MY) 2020/21 will forecast up 22 million tons in comparison with 2019/2020 MY, reaching 188 million tons (raw value), due to higher production in Brazil, India, and Thailand [1]. These countries alternate sugarcane uses for obtaining sugar, ethanol and other products, generating near to 152 million tons of residues.

It is estimated that only one-third of the energy potential of sugarcane is derived from its juice [2, 3], which has been efficiently used in the production of sugar and first-generation ethanol. The remainder of the energy potential is associated with the sugarcane bagasse and straw, which represent approximately two thirds of the crop energy potential [2–4].

In this context, Brazil is a world leader in renewable electricity generation and increased the focus on electricity generation from sugarcane biomass over the last few decades, driven especially by the increasing price of electricity sold to the grid, public–private initiatives and specific policies to encourage sales of surplus electricity [5].

Initially, electricity was generated only to meet the self-consumption supply of the sugarcane mills, but with a modernization of cogeneration systems and the growing use of bagasse and, in some cases, straw, many sugarcane mills have become net exporters of electricity [6]. In fact, cogeneration has become one of the most efficient technologies for the realistic use of primary fuel to produce electricity and heat [7]. Thus, since 2013, surplus electricity offered to the grid by the sugarcane sector has been greater than that used for self-consumption (i.e., in 2019, about 61% exported to the grid versus 39% for self-consumption; **Figure 1**).

However, although Brazilian mills already recover energy from bagasse trough cogeneration systems, there are other residues from sugar and ethanol production process that can also be valorized by recovering their energy: straw, vinasse, and filter cake. They can be used in energy conversion processes, such as the production of biofuels or in cogeneration processes, and also as a source of carbohydrates from other biomolecules of commercial interest, such as second generation ethanol, xylitol, enzymes, organic acids, proteins, and other bioproducts of industrial interest [8].

Nowadays, bagasse and, sometimes, straw are used for energy recovery through combustion, while filter cake and vinasse are commonly used for soil fertilization. The use of these residues as energy sources depends on their chemical and physical properties. In this context, this chapter has the main objective of presenting an updated literature review for these properties of the main sugar and alcohol industry's by-products, considering their use in energy recovery processes.

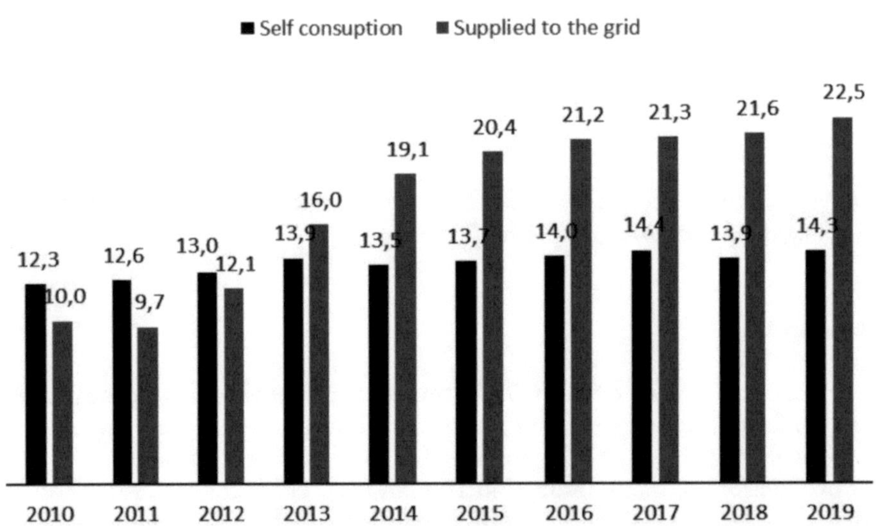

Figure 1.
Bioelectricity generation from sugarcane residues (terawatt hour – TWh) in Brazil from 2010 to 2019 [6].

2. Sugar and alcohol production process and its by-products

The main residues from sugar and alcohol production are straw, bagasse, vinasse and filter cake. A simplified scheme of sugar and ethanol production from sugarcane is shown in **Figure 2**, where the generation of each residue can be seen.

The straw, also called trash, is an agricultural waste composed by tops and leaves of the sugarcane. Sugarcane straw comes from its mechanized harvest, as shown in **Figure 2**, and its production (on a mass basis) depends on some factors such as the harvesting system, the height of the tips, the sugarcane variety, the age of the crop (cutting stage) besides soil and climate conditions, among others [9]. Some authors reported that straw production varies between 10 and 18 ton/ha (dry basis) and the ratio straw (dry basis)/stalk (wet basis) between 11 and 17% [4].

The straw used to be burned during the manual harvest, but, since the 2010/2011 harvest, the mechanization started to be disseminated. In this process, the green and dry leaves and the tips of the sugarcane are cut and mixed in the extractor. With this, part of the straw is left at the field, for soil protection, and part of it is carried to the mills, together with the sugarcane with levels of impurity that vary according to the harvesting system used [10].

Several studies indicate that sugarcane fields contain an average of 8–30 ton·ha^{-1} dry mass of straw and its production varies according to crop variety, vegetative stage, edaphoclimatic conditions and management practices [5]. Straw has similar properties

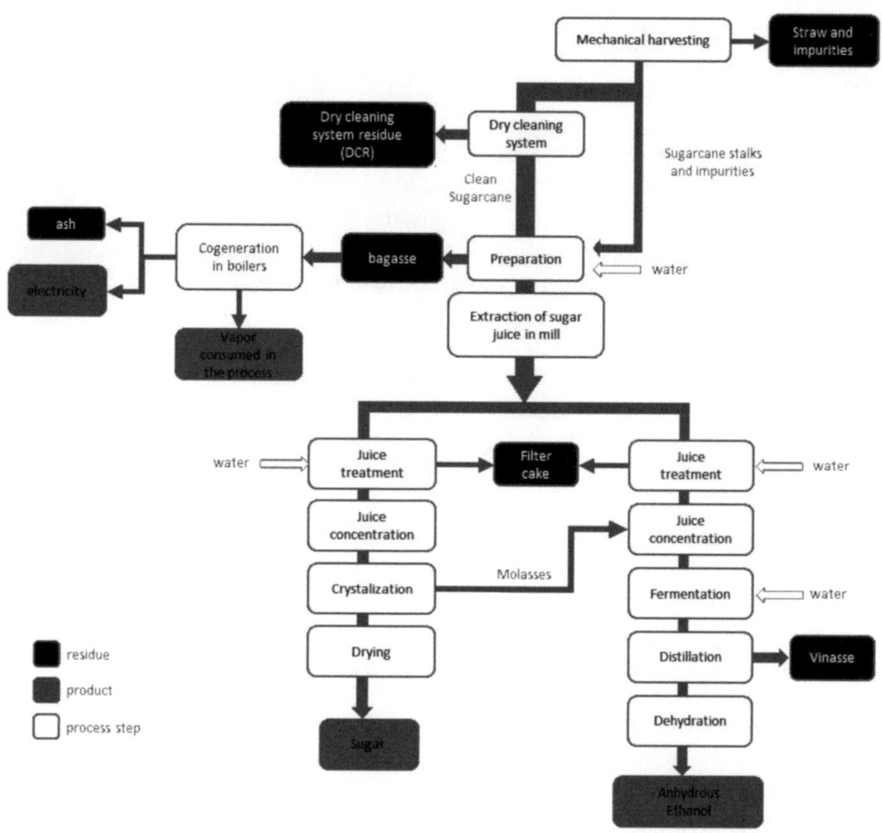

Figure 2.
Simplified scheme of sugar and ethanol production from sugarcane.

to bagasse, which makes it a good fuel to supplement bagasse for surplus power generation at mill through the burn system. However, considering that straw is also necessary in field for soil protection, it is only partially available for energy recovery. The amount of straw left in the field in relation to the total production of sugarcane taken to the mill can vary from 10 to 60% in Colombia and from 20 to 35% in South Africa [9].

According to [11], an average of 140 kg of straw is generated for each ton of harvested cane (dry basis). This represents nearly the same amount of bagasse available at the end of the juice extraction process. Some authors estimated the total electricity surplus when 40–50% of the straw available in the field is used as additional fuel to bagasse and showed results between 130 and 185 kWh per ton of cane [12, 13].

In the work of [9], it was estimated that the amount of straw left in the field per sugarcane hectare is 10 tons/ha in India. In other hand, the same author presents that 39 tons of straw/sugarcane ha are left on the soil and its productivity is of 81.49 t/ha in the region of the Rio Grande Valley, Texas in the United States. The authors report some benefits of keeping sugarcane residue in the field, such as:

- Protection of the wind surface of the soil against erosion caused by rain and hair;

- Reduction of variations in soil temperature, as the soil is protected from the direct action of solar radiation;

- Increased biological activity in the soil;

- More water available, due to reduced water evaporation from the soil surface;

- Weed control.

After the harvesting, the sugarcane cleaning is required, which can be carried by dry system or by cane washing. The dry-cleaning system residue, or simply DCR, is produced from the dry cleaning of the sugarcane right after its mechanized harvest in the field, as shown in **Figure 2**. That is, the sugarcane goes through a cleaning process without the use of water (sugarcane washing) to remove soil and impurities (leaves and straw).

The cleaning system has the advantage of reducing water consumption by plants, however, as its use is not yet very diffused, the cleaning residue properties are still unknown and, in most cases, it ends up being discarded.

As it can be seen in **Figure 2**, after sugarcane cleaning, the next step is to extract the juice through mills or diffusers, from which the bagasse, another by-product of the process, comes from.

The bagasse is the fiber that results from juice extraction process in the mills. It is estimated that 135 kg of bagasse are generated for each ton of clean cane [11]. According to [10, 14], one ton of harvested sugar cane (wet basis) originates about 250–270 kg of bagasse containing 50% moisture, 48% fibers and 2% soluble solids. Bagasse properties can vary according to the process for juice extraction [15].

Sugar and alcohol plants use the bagasse resulting from the ethanol production process in cogeneration systems, converting their chemical energy into thermal and electrical energy for their own plant. Sugarcane facilities currently use a cogeneration system based on the Rankine cycle, in which sugarcane bagasse is burnt in a boiler, producing steam that is expanded in turbines coupled with electric generators; turbines exhaust steam is used as thermal energy source for the various unit operations of the sugar and ethanol production process. Most facilities use only back-pressure steam turbines, which limit the amount of fuel that can be burnt to supply the steam demand of the process [16].

After sugarcane juice extraction, the next steps are to filtrate and to treat it, from which filter cake, or press mud, comes out as residue (**Figure 2**). It is a compound from the clarification and filtration of the juice extracted from the mills in the rotating filters, consisting in a solid fibrous residue that represents about 3–4% (w/w) of the crushed sugar cane [17].

Sugarcane filter cake is a semi-solid material composed of fiber, crude protein, sugar, wax, fat, and ash [18, 19]. It is a rich source of phosphorus and organic matter, with a moisture content, reasons why it has been used as a complete or partial substitute for mineral fertilizers in sugar cane or other crops cultivation. This using has been widespread in several countries, including Brazil, India, Australia, Cuba, Pakistan, Taiwan, South Africa, and Argentina [20].

Vinasse is produced from the distillation of wine during the hydrated ethanol production process. It is an aqueous effluent and a problem to the sector due to the large quantities produced and its potential effects as an environmental pollutant. It is largely composed of water, organic matter, and mineral elements. The environmental damage caused by discarding it into the soil or running waters was an incentive to studies aiming to find alternative, economic applications for this residue. Results from such studies indicate that, when properly applied, vinasse contributes to improvements in soil quality and agricultural productivity [20], while others, show that sue to vinasse and the filter-cake residues show considerable K and P concentrations, they are used as fertilizer [21–24].

According to [25], the vinasse generation is directly proportional to the alcohol production, obtaining between 11 and 15 liters of vinasse for each liters of ethanol [25, 26]. The study of [27, 28] state that, for autonomous distillery, approximately 910 liters of vinasse are obtained for each ton of processed sugarcane with 250 kg of bagasse and 30 kg of filter cake.

Among the residues from the alcohol production process, vinasse is considered to have the greatest polluting potential, since it has a solids content of around 7%, of which 75% is organic and biodegradable, in addition to having high COD (Chemical Oxygen Demand) and BOD (Biological Oxygen Demand) [29]. According to [30], the vinasse BOD and COD can varies, respectively, from 6,000 to 25,000 and 15,000 to 65,000 mgO_2/L, depending on whether it is derived directly from the juice or from the cane syrup. In addition, this residue has a corrosive character due to the fact that it is a buffered solution with a pH around 4.3 obtained at high temperature [29].

There are processes by which the polluting potential of vinasse and filter cake is minimized. Nowadays, both are directly applied as fertilizers, reducing the demand for irrigation and mining in the cultivation of sugarcane. However, according to [31], the anaerobic biodigestion process can be used in the treatment of these residue. This process has the advantage of treating the vinasse and the filter cake, keeping their fertilizing properties, while produces the biogas, composed of a mixture of methane gas (CH_4) and carbon dioxide (CO_2) in different proportions, depending on the substrate contained in wastewater. Biogas is an environmentally friendly alternative fuel that is, thus the liquid waste vinasse is utilized commercially [32].

3. Physicochemical characterization of sugarcane industry by-products and their use in energy processes (straw, bagasse, filter cake, vinasse)

3.1 Heating value

There are two variations for heating value. The higher heating value (HHV) which refers to the heat released during combustion per unit mass of the fuel, considering that the water formed in the combustion is in a liquid phase, in which the temperature of the gases and water produced is equal to the temperature of fuel before combustion. The lower heating value (LHV) is the heat released during combustion per unit mass of the fuel, considering that the water formed during combustion is in a gaseous phase, in which the temperature of the gases and water produced is the same as that of the fuel before combustion [33].

The HHV can be determined experimentally, using the methodology described in ASTM D2015–00 Standard test method for gross heating value of coal and coke by adiabatic bomb calorimeter by means of a calorimetric pump [34].

Normally, the heating value of solid fuels is determined in a calorimetric pump, that is, the higher heating value is calculated on a dry basis (zero moisture content), from which the lower heating value can be calculated, which is used, for example, in the design of boilers.

The study of [33] present an equation for calculating the lower heating value on a wet basis, from the higher heating value on a dry basis, both in $MJ{\cdot}kg^{-1}$, according to Eq. (1):

$$LHV = HHV \cdot \left(1 - \frac{H_2O}{100}\right) - 2.444 \cdot \frac{H_2O}{100} - 2.444 \cdot \frac{h}{100} \cdot 8.936 \cdot \left(1 - \frac{H_2O}{100}\right) \quad (1)$$

Where:

LHV = lower heating value on a wet basis ($MJ{\cdot}kg^{-1}$);

HHV = higher heating value on a dry basis ($MJ{\cdot}kg^{-1}$);

H_2O = moisture content of the fuel on a wet basis (% mass);

h = Hydrogen concentration on a dry basis (% mass).

The first term of the equation converts the HHV from dry to wet basis, the second term corresponds to the latent heat of water vaporization contained in the biomass (the latent heat of water vaporization at 25°C and constant pressure is 2.444 $MJ{\cdot}kg^{-1}$). Finally, the third term refers to the vaporization of the water produced when the hydrogen contained in the biomass is combusted.

The author in [33] also refer to other empirical equations for calculating higher heating value, as shown in Eq. (2), described by [35].

$$\begin{aligned} HHV = {} & 0.3491 * X_C + 1.1783 * X_H + 0.1005 * X_s - 0.0151 * X_N \\ & - 0.1034 * X_O - 0.0211 * X_{ash} \end{aligned} \quad (2)$$

Where:

X_i: dry mass fraction of Carbon (C), Hydrogen (H), Sulfur (S), Nitrogen (N), Oxygen (O) and Ash.

HHV = higher heating value on a dry basis ($MJ{\cdot}kg^{-1}$);

Through empirical Eq. (2), it is observed that the levels of carbon, hydrogen and sulfur will contribute to the higher heating value of the material, while the levels of nitrogen, ash and oxygen contribute negatively. Although it reduces the higher heating value of solid fuel, organic oxygen is released during thermal decomposition and thus supplies part of the oxygen needed for combustion reactions, thus decreasing the amount of air required in the process.

In **Tables 1** and **2**, data of the heating value of bagasse and sugarcane straw are shown, both lower and higher, respectively, reported in several studies carried out with samples from different countries of origin.

From the data in **Table 1**, it is observed that the values for the higher heating value of bagasse vary from 15.98 to 19.17 $MJ{\cdot}kg^{-1}$, while for straw is about 17 $MJ{\cdot}kg^{-1}$. It is interesting to highlight the differentiation made by [5] in relation to the thermochemical properties relevant to the different constituent parts of the straw.

HHV (MJ·kg^{-1})	LHV (MJ·kg^{-1})	Base	Author
15.98	14.67	Wet base 9.22%	[36]
17.0	—	—	[37]
18.65	—	—	[38]
18.1	—	Dry base	[9]
16.88	—	Dry base	[39]
18.89	17.32	—	[40]
18.90	—	—	[15]
18.61	—	Dry base	[41]
17.23	—	Dry base	[42]
18.59	—	Dry Base	[43]
18.43	—	Dry Base	[43][a]

[a]*Bagasse resultant from a process with dry cleaning system of sugarcane.*

Table 1.
Literature review of the heating value of sugarcane bagasse.

Biomass	HHV (MJ·kg^{-1})	LHV (MJ·kg^{-1})	Author
Green leaves	17.4	—	[9]
Dry leaves	17.4	—	[9]
Tips	16.4	—	[9]
Straw	17.74	16.50	[44]
Dry leaves	16.0	14.8	[41]
Straw	16.347	—	[45]
Straw[a]	16.54	—	[43]
Straw	17.1	—	[10]

[a]*Straw resultant from a process with dry cleaning system of sugarcane.*

Table 2.
Literature review of the heating value on dry basis of sugarcane straw.

In the work of [36], states to the fact that the viability of using sugarcane residues as a fuel in cogeneration processes is directly related to their moisture content, since the heating value is intrinsically linked to fuel's elemental composition and moisture.

As a comparison criterion for this important thermochemical property for the biomass combustion process, the heating value of the wood can be taken as a reference. For this, [46] shows that this traditional biomass can present HHV very close to that found for bagasse and straw, being 18.49 MJ·kg^{-1} for sawdust from Eucalyptus sp.

On the other hand, [37] shows that coal has a calorific value of 29 MJ·kg^{-1}, that is, a heating value about 49% higher than that found for bagasse and straw, mainly due to its greater constituent carbon content.

According to the **Table 3**, vinasse high heating value fluctuates from 12.7 MJ·kg^{-1} to 15.07 MJ·kg^{-1}, when moisture is around 4% and, takes a value of 6.4 MJ·kg^{-1} with 68% of moisture [47, 51]. Hence, vinasse high heating value decreases notably with the increase in moisture content, which must be considered when a thermochemical route is used for energy recovery [52].

HHV ($MJ \cdot kg^{-1}$)	LHV ($MJ \cdot kg^{-1}$)	Author
13.59	—	[47]
14.40	—	[48]
6.4	4.5	[49][a]
13.0	14.0	[50]
12.7	12.6	[51]

[a]*When moisture around 68%.*

Table 3.
Literature review of the heating value on dry basis of sugarcane vinasse.

As showed in **Table 4**, sugarcane Filter-cake have around 17 $MJ \cdot kg^{-1}$ of high heating value, which is proximate to the results reported for straw and bagasse in **Tables 1** and **2**. This indicates a high energy potential in filter cake, which could be related to a high content of proteins, sugar and fibers, giving a possibility to use this by-product for energy as with other sugarcane and woody biomasses which is normally used for thermochemical processes [56, 57].

Taking into account the energy potential of the sugarcane industry residues, there is a possibility to use it as the coal. One favorable way to increase residues potential could be to use a blend of them. Residues as the vinasse and filter cake, which have a high availability, would present more heating value compensate by add straw and bagasse [51].

3.2 Ultimate analysis

An ultimate analysis of biomass offers the contents of Carbon, Nitrogen, Hydrogen, Oxygen and Sulfur, which together with moisture and ash, are the main components of biomass. With the results of this analysis, it is possible to estimate the amount of products that will generated in combustion or gasification, as well as the amount of oxidant needed. The nitrogen content is useful for assessing the amount of nitrogen oxide (NO_x) transport in combustion or ammonium (NH_4) in gasification. The sulfur content is necessary to calculate the possible dates of sulfur dioxide (SO_2) or hydrogen sulphide (SH_2) and with chlorine, to assess corrosion of the equipment [54].

The elementary analyses of sugarcane bagasse and straw found in the literature of recent years are summarized in **Tables 3** and **4**.

Regarding the properties of wood, the sawdust of Eucalyptus sp., it has a carbon content of 46.80%, oxygen of 46.61% and hydrogen oxygen of 6.59% [46]. Therefore, it can be noted that bagasse and cane straw have an ultimate composition similar to wood, ancient biomass used in thermochemical processes.

In addition, bagasse and sugarcane straw have a 39% lower carbon content compared to coal [37]. Even so, it can be said that these by-products of the sugar and alcohol sector have combustible characteristics, since carbon and hydrogen are responsible for more than 50% of their composition, which will undergo oxidation in the presence of air during combustion, releasing heat.

In addition, from the results showed in **Table 5** and form other authors [60–63], bagasse and straw have a high oxygen content (from 39 to 50%), when compared to coal, which may have 7.4% oxygen content, according to [37], which implies low heating value of bagasse and straw compared to coal. However, the high content of constituent oxygen reduces the amount of oxygen required in the combustion process, since organic oxygen is released during thermal decomposition and provides part of the oxygen needed for combustion reactions (**Tables 6** and 7) [33].

HHV (MJ·kg^{-1})	LHV (MJ·kg^{-1})	Author
17.05	14.6	[53]
17.0	—	[54]
17.1	—	[55]
14.9	13.8	[50][a]
8.57	7.79	[56]

[a]*Wet basis.*

Table 4.
Literature review of the heating value on dry basis of sugarcane filter cake.

C (%)	H (%)	O (%)	N (%)	S (%)	Cl (%)	Author
44.8	5.40	39.60	0.4	0.01	—	[58]
44.8	5.4	38.1	0.4	0.03	0.02	[37]
50.3	6.3	43.1	0.3	0.07	—	[38]
44.6	5.8	44.5	0.6	0.1	0.02	[9]
44.31	5.73	49.11	0.63	<0.1	0.13	[39]
43.6	5.52	50.63	0.25	0.07	—	[59]
45.48	5.70	45.21	0.40	0.06	—	[15]

Table 5.
Ultimate composition of sugarcane bagasse reported by different authors.

Biomass	C (%)	H (%)	O (%)	N (%)	S (%)	Cl (%)	Author
Green leaves	45.7	6.2	42.8	1.0	0.1	0.4	[9]
Dry leaves	46.2	6.2	43.0	0.5	0.1	0.1	[9]
Tips	43.9	6.1	44.0	0.8	0.1	0.7	[9]
Straw	44.7	5.8	—	0.45	0.08	—	[44]
Straw	43.42	5.71	49.64	1.23	—	—	[10]
Straw	42.5	6.02	50.2	0.6	0.24	0.44	[10]

Table 6.
Ultimate composition of sugarcane straw reported by different authors.

Considering that sugarcane vinasse is obtained in the final stages of bioethanol production, their composition is highly heterogeneous, which may explain why the nitrogen content is approximately three times higher than in straw and bagasse. The high Carbon and Oxygen content verify residual sugars and acids from the sugarcane process. On the other hand, sulfur content can be explained by the presence of SO_2 residue from sulphitation in cane juice treatment [57].

According to **Table 8**, the high content of oxygen and carbon suggests the presence of organic groups characteristic of lignocellulosic materials, which may indicate their use as solid biofuel or for obtaining bio-oil [68]. In addition, as an important part of the filter cake is obtained in the sulfite and clarification processes, a high sulfur content may result when compared to other sugar cane residues. Although the presence of Nitrogen and Sulfur in most of the residues studied is low, when compared to other biomasses, they are undesirable because they can reduce

C (%)	H (%)	O (%)	N (%)	S (%)	Cl (%)	Author
41.2	5,0	20.8	5.0	5.0	—	[49]
32.9	4.5	36.4	1.0	2.7	6.2	[50]
31.87	6.13	28.26	1.69	2.32	—	[51]
39.7	8.6	—	1.65	0.12	—	[64]

Table 7.
Ultimate composition of sugarcane vinasse in dry basis reported by different authors.

C (%)	H (%)	O (%)	N (%)	S (%)	Cl (%)	Author
32.5	2.2	—	2.2	—	—	[65][a]
42.9	5.3	24.99	1.8	3.1	—	[53]
33.73	3.92	—	2.36	0	—	[54]
37.1	—	—	2.3	—	—	[66]
29.6	3.9	46.0	0.9	0.5	0.3	[50]
21.5	3.33	18.34	0.81	0.13	—	[56]
34.4	—	—	2.07	9.93	—	[67][b]

[a]*Wet basis.*
[b]*Dry basis.*

Table 8.
Ultimate composition of sugarcane filter cake reported by different authors.

the calorific value, at the same time as they decrease combustion efficiency by promoting the formation of nitrogen oxides (NO_x) and sulfur oxides (SO_x) [51, 57].

It is also important to take into account that the chemical composition of sugarcane residues depends on the locality, cane variety, land conditions, nutrients applied to the field, milling efficiency and method of clarification [54, 57].

3.3 Proximate analysis

The proximate analysis establishes the contents of moisture, ash, volatile material and fixed carbon of the fuels, with which it is possible to estimate the behavior of the biomass during the combustion process.

The moisture content can determine the most suitable energy conversion process for biomass, since for materials with moisture content greater than 50%, biological routes are generally applied, such as anaerobic decomposition, while for biomasses with moisture content of up to 50% (wet basis), thermochemical processes are more used [69–72].

The fixed carbon indicates the carbon content that still remains in the biomass used as fuel, after the volatilization of the volatile compounds. An indicative relationship can be established between these two levels, in which materials with a higher volatile content in relation to the fixed carbon content will be combustible with greater ignition ease. Finally, ash is the inert material that results from combustion, so the higher the ash content, the less energy is available in the fuel [69].

The moisture content of straw depends, in general, on the time that this material is deposited in the field. At the time of harvest, the straw can contain up to 50% moisture (w.b.). This moisture content can drop to 30% (w.b.) in 2 to 3 days and to 15% (w.b.) in two weeks, which can represent an energy gain in its burning [9]. For bagasse, the moisture content in the juice extraction outlet at the plants, whether by milling or diffusion, is in the range of 47 to 52% (w.b.) [33].

The values of the contents of fixed carbon, volatile matter and ash for bagasse and straw, determined by different authors, are shown in **Tables 9** and **10**.

The proximate analysis data for the bagasse and straw are somewhat similar in terms of variation: the fixed carbon is within the range of 6.9–18% and the volatile material is 73–89%. These variations in values may be related to the varieties of sugar cane, as well as the presence of impurities.

Fixed carbon terminology refers to the fuel fraction of coal or any bituminous material after removal of moisture, volatile and ash content. Or it can also be defined as the elemental carbon content constituting coal and bituminous materials added to the fraction of carbonaceous residue formed under the rate of heating of the material [75]. In the case of the methodology described by ASTM E872–82, this heating is done at 950°C for 7 minutes [76].

However, according to [75], biomass does not have elemental carbon, unlike coal, since its carbon fixation occurs through photosynthetic fixation to form the main macromolecules that make up biomass, such as carbohydrates, lipids and proteins. Thus, the term fixed carbon is not adequately used in biomass and its corresponding fraction refers more to a content of pyrolytic carbon (carbon formed at a temperature of approximately 100°C). In addition, the pyrolytic carbon content will directly depend on the temperature rate used for the determination of volatile material in the biomass, as already described.

The content of volatile materials, in turn, refers to the amount of material that will detach from the fuel in gaseous form, undergoing combustion first. Thus, the higher the content of volatile materials in biomass, the better its combustion. In relation to this, it is observed that both straw and bagasse have a high content of

Fixed carbon (%)	Volatile matter (%)	Ash (%)	Author
11.95	85.61	2.44	[73]
14.9	73.8	11.3	[51]
18.0	79.9	2.2	[20]
13.1	86.0	0.9	[54]
9.3	88.7	2.0	[33]
—	—	3.1	[74]
—	—	2.4	[74]
12.4	81.8	5.8	[74]

Table 9.
Proximate analysis for sugarcane bagasse on a dry basis reported by different studies.

Biomass	Fixed carbon (%)	volatile matter (%)	Ash (%)	Moisture (%)	Author
Green leaves	15.7	80.6	3.7	67.7*	[9]
Dry leaves	11.6	84.5	2.7	13.5*	[9]
Tips	16.4	79.3	4.3	82.3*	[9]
Straw	6.9	81.55	11.57	9.92	[44]
Straw	17.46	78.64	4.32	—	[45]
Straw	10.1	82.5	7.5	—	[10]

**Moisture of fresh samples.*

Table 10.
Proximate analysis for sugarcane straw reported by different studies.

volatile materials (about 80%), when compared to coal (34%), which indicates that they are easily combustible fuels [36, 37].

On the other hand, wood residues present volatile material contents of 88.27% (sawdust from Eucalyptus sp.) And 80.73% (sawdust from Pinus sp). Therefore, it is noted that the residues of the sugar and alcohol sector show similar behavior to wood during direct combustion, considering the high content of volatiles [46].

Regarding the ash content, it is noted that the sugarcane straw has a higher ash content compared to the bagasse, since for the bagasse the ash content is between 0.9 and 11.57% and for the straw this variation ranges from 2.7% to 11.3%.

The analysis of the ash content of the sugarcane residues at the sugar and alcohol plant is of paramount importance as regards the use of these materials in thermochemical conversion processes, considering that the lower the ash content, the less the problems in the boilers such as scale, deposits and corrosion.

In this sense, the ash values found for bagasse and straw are close to those found for wood sawdust, which may vary from 0.46% for sawdust from Eucalyptus sp. to 7.88% for sawdust from Pinus sp. Regarding mineral coal, both straw and bagasse have much lower ash contents, which can reach a difference of 13.8%, according to [37].

Tables 11 and **12** reports the proximate composition of vinasse and filter cake, respectively, according to different studies from literature.

Considering **Tables 11** and **12**, information for filter-cake and vinasse moisture fluctuate according to the sample utilized for the study, but normally the sample need to be dried. *In natura* conditions, these by-products have a moisture around 90%, which may imply low yields in thermochemical transformation processes, but is an important design parameter for assessing the need for a previous drying step and estimating the energy consumption involved [72].

Analyzing the volatile material, it can be observed that the vinasse and the filter cake have a lower content, compared to the bagasse and straw, indicating a lower reactivity and the need of high temperatures to achieve its thermal degradation. Considering the diversity in the composition of vinasse and filter cake, the content of volatile materials probably corresponds to the evaporation of low molecular weight compounds such as lactic acid, phenolic compounds and, finally, to the partial release of combustible (CxHy gas, CO and H_2) and non-combustible (CO_2, SO_x, NO_x H_2O) products as well as straw and bagasse. Thus, the organic composition (Volatile Material) of vinasse and filter-cake and the predominance of inorganic composition (Ashes), reduces the energy that can be released [51, 52, 73, 78].

In **Table 11**, fixed carbon presented for vinasse may indicates longer combustion and a higher thermochemical conversion rate compared with filter-cake, as showed in **Table 12**. Considering the ash content, vinasse has a value in the range of feedstock material normally used for pyrolysis and gasification processes [50]. On the other hand, filter-cake present highest ash content than vinasse, which may adversely affect

Fixed carbon (%)	Volatile matter (%)	Ash (%)	Moisture(%)	Author
18.95	69.31	11.73	—	[74]
—	32,3	9,7	58	[49]
—	—	34.1	—	[50]
3.55	63.06	29.35	4.05	[51]
12.18	61.66	20.56	5.6	[56]

Table 11.
Proximate analysis for sugarcane vinasse reported by different studies.

Fixed carbon (%)	Volatile matter (%)	Ash (%)	Moisture(%)	Author
25.95	56	15.55	—	[53]
—	61.2	25.9	3.52	[54]
—	—	23.32	73.13	[77]
—	—	42.6	—	[50]
—	80.8	19.2	73.3	[75]
3.45	40.67	45.83	2.06	[56]
—	—	20.6	10.5	[67]

Table 12.
Proximate analysis for sugarcane filter cake reported by different studies.

the high heating value, important in a thermochemical process, but this high mineral measure and possibly sugar and protein content, according ultimate analysis, indicates filter cake can be used in the biogas production [79]. Likewise, these ash contents in sugarcane waste increase the sintering potential and the possibility of fouling and corrosion in combustion reactors or boilers, as well as operating costs [80].

4. Conclusions

Sugarcane industry residues presents a feasibility to be used in energy conversion process, as indicated by the physicochemical properties showed in this chapter.

The high heating value is the principal parameter to analyze the energy that can be liberated by a biomass in a thermochemical process. Most of the uncommon sugarcane residues heating values were proximate, which if compared with woody biomass or bagasse, normally used in a combustion process, allows to demonstrate the potential for energy recovery in the sugarcane industry.

However, vinasse and filter cake present high moisture and ash contents, which indicates that these residues must present better yield in biochemical processes of energy conversion instead of the thermochemical ones. The high ash concentration is related to a good performance of the biofertilizer obtained after the biogas production, which can be used for fertilizing sugarcane crops. The high concentration of carbon, in appropriated ratio with nitrogen, indicates that good yields of methane gas can be obtained.

On the other hand, the production of biogas requires the construction of an appropriate system for anaerobic biodigestion, which requires investment by the plants in new facilities and processes. In this case, the joint use of residues in thermochemical processes can be considered, causing a compensatory effect of the straw and bagasse with the properties of the vinasse and filter cake, mainly when it is required to improve the negative effects related to the ultimate and proximate composition, to increase the energy potential of the waste.

Acknowledgements

The authors acknowledge to UFABC, CAPES, CNPq and FAPESP (Processes number and 2011/51902-9) for the financial support. Also, the authors would like to thank the UFABC Multi-user Experimental Center (CEM) for support in experimental activities. The opinions, hypotheses and conclusions or recommendations expressed in this material are of authors responsibility and do not necessarily reflect FAPESP vision.

Appendices and nomenclature

MY	marketing year
TWh	terawatt hour
DCR	dry cleaning system residue
COD	chemical oxygen demand
BOD	biological oxygen demand
LHV	lower heating value
HHV	higher heating value
w.b.	wet basis

Author details

Julia M. de O. Camargo, Jhuliana Marcela Gallego Ríos*, Graziella C. Antonio and Juliana T.C. Leite
Federal University of ABC (Universidade Federal do ABC), Santo André, Brazil

*Address all correspondence to: juliana.toneli@ufabc.edu.br

IntechOpen

© 2021 The Author(s). Licensee IntechOpen. This chapter is distributed under the terms of the Creative Commons Attribution License (http://creativecommons.org/licenses/by/3.0), which permits unrestricted use, distribution, and reproduction in any medium, provided the original work is properly cited.

References

[1] United States Department of Agriculture (USDA). Foreign Agricultural Service. Sugar: World Markets and Trade. [Internet]. 2020. Available from: https://apps.fas.usda.gov/psdonline/circulars/Sugar.pdf

[2] Alonso-Pippo, W. A.; Luengo, C. A.; Alberteriz, L. A. M.; Garzone, C.; Cornacchia, G. Energy Recovery from Sugarcane-Trash in the Light of 2nd Generation Biofuels. Part 1: Current Situation and Environmental Aspects. Waste and Biomass Valorization. 2011. 2: 1-16.

[3] Alonso-Pippo, W. A.; Luengo, C. A.; Koehlinger, J.; Garzone, P. Cornacchia, G. Sugarcane energy use: The Cuban case. Energy Policy. 2008. V. 36, N. 6, p. 2163-2181

[4] Leal, M. R. L. V., Galdos, M. V., Scarpare, F. V., Seabra, J. E. A., Walter, A., & Oliveira, C. O. F. Sugarcane straw availability, quality, recovery and energy use: A literature review. Biomass and Bioenergy. 2013; 53: 11-19. DOI:10.1016/j.biombioe.2013.03.007

[5] Carvalho J. L. N., Nogueirol R. C., Menandro L. M. S., Bordonal R. de O, Borges C. D., Cantarella H., Franco H. C. J. Agronomic and environmental implications of sugarcane straw removal: a major review. GCB Bioenergy. 2016; 9(7):1181-1195. DOI:10.1111/gcbb.12410

[6] Union of the Sugarcane industries UNICA. A bioeletricidade da cana em numeros. [Internet]. 2020. Available from: https://observatoriodacana.com.br/listagem.php?idMn=134 [Accessed: 2020-11-04]

[7] Guerra, J. P. M., Coleta, J. R., Arruda, L. C. M., Silva, G. A., & Kulay, L. Comparative analysis of electricity cogeneration scenarios in sugarcane production by LCA. The International Journal of Life Cycle Assessment. 2014; 19(4): 814-825. DOI:10.1007/s11367-014-0702-9

[8] Chandel A. K., Da Silva S. S., Carvalho W., Singh O. V. Sugarcane bagasse and leaves: foreseeable biomass of biofuel and bio-products. Journal of Chemical Technology and Biotechnology. 2011; 87 (1): 11-20. Available from: https://doi.org/10.1002/jctb.2742

[9] Hassuani S. J., Leal M. R. L. V., de Carvalho Macedo, I. Biomass power generation: sugar cane bagasse and trash. In: CTC Eds. 2005. 217 p.

[10] Bizzo W.A., Lenço P.C., Carvalho D.J., Veiga J.P.S. The generation of residual biomass during the production of bio-ethanol from sugarcane, its characterization and its use in energy production. Renewable and Sustainable Energy Reviews. 2014; 29:589-603.

[11] Macedo IC, Seabra JEA, Silva JEAR (2008) Green house gases emissions in the production and use of ethanol from sugarcane in Brazil: the 2005/2006 averages and a prediction for 2020. Biomass Bioenerg 32:582-595. doi:10.1016/j.biombioe.2007.12.006

[12] Seabra JEA, Macedo IC. Comparative analysis for power generation and ethanol production from sugarcane residual biomass in Brazil. Energ Policy 2011;39:421e8

[13] Dias MOS, Cunha MP, Jesus CDF, Rocha GJM, Pradella JGC, Rossell CEV, et al. Second generation ethanol in Brazil: can it compete with electricity production? Bioresour Technol 2011;102:8964e71

[14] CORTEZ, L.; MAGALHAES, P.; HAPPI, J. Principais subprodutos da agroindústria canavieira e sua

valorização. Sociedade Brasileira de Planejamento Energético Revista Brasileira de Energia, v. 2, n. 2, p. 1-17, 1992.

[15] Lenço P. C. Caracterização do bagaço de cana-de-açúcar para geração de energia. [thesis]. Faculdade de Engenharia Mecânica: Universidade Estadual de Campinas, Campinas;2010.

[16] Dias M. O. de S., Maciel Filho R., Mantelatto P. E., Cavalett O., Rossell C. E. V., Bonomi A., Leal M. R. L. V. Sugarcane processing for ethanol and sugar in Brazil. Environmental Development. 2015; 15: 35-51. DOI: 10.1016/j.envdev.2015.03.004

[17] Baez-Smith, C. Production of Bioenergy Using Filter Cake Mud in Sugar Cane Mill Factories; Sugar Processing Research Institute, Inc. (SPRI) : New Orleans, LA, 2008.

[18] Lopez González, L. M.; Pereda Reyes, I.; Dewulf, J.; Budde, J.; Heiermann, M.; Vervaeren, H. Effect of liquid hot water pre-treatment on sugarcane press mud methane yield. Bioresour. Technol. 2014, 169 (0), 284-290.

[19] Rouf, M. A.; Bajpai, P. K.; Jotshi, C. K. Optimization of Biogas Generation from Press Mud in Batch Reactor. Bangladesh J. Sci. Ind. Res. 2011, 45 (4), 371-376.

[20] Prado R. de M., Caione G., Campos C. N. S. Filter Cake and Vinasse as Fertilizers Contributing to Conservation Agriculture. Applied and Environmental Soil Science. 2013; 1-8. DOI:10.1155/2013/581984

[21] Marisa S. Crespi; Quézia V. Martins; Sonia de Almeida; Hernane S. Barud; Marcelo Kobelnik; Clóvis A. Ribeiro. Characterization and thermal behavior of residues from industrial sugarcane processing. 2011, 106(3): 753-757. DOI:10.1007/s10973-011-1397-9.

[22] Da Mota, R. P., de Camargo, R., Lemes, E. M., Lana, R. M. Q., de Almeida, R. F., & de Moraes, E. R. Biosolid and sugarcane filter cake in the composition of organomineral fertilizer on soybean responses. International Journal of Recycling of Organic Waste in Agriculture. 2018. DOI:10.1007/s40093-018-0237-3

[23] Ochoa George, P. A., Eras, J. J. C., Gutierrez, A. S., Hens, L., & Vandecasteele, C. Residue from Sugarcane Juice Filtration (Filter Cake): Energy Use at the Sugar Factory. Waste and Biomass Valorization. 2010, 1(4): 407-413. DOI:10.1007/s12649-010-9046-2

[24] Borges, B. M. M. N., Abdala, D. B., Souza, M. F. de, Viglio, L. M., Coelho, M. J. A., Pavinato, P. S., & Franco, H. C. J. Organomineral phosphate fertilizer from sugarcane byproduct and its effects on soil phosphorus availability and sugarcane yield. Geoderma. 2019, 339: 20-30. DOI:10.1016/j.geoderma.2018.12.036

[25] Molina Betancourt, A. C. Biodegradación anaeróbica de vinaza generada en la destilería Soderal y aprovechamiento energético del biogás como medio para bajar el impacto de gases de efecto invernadero [Master's thesis]. Universidad Politécnica Salesiana Sede Guayaquil; 2012.

[26] FREIRE, W. J.; CORTEZ, L. A. B. Vinhaça de cana-de-açúcar. Guaíba: Agropecuária, 2000. 203p.

[27] Orlando Filho J. Nutrição e Adubação da cana-de-açúcar no Brasil. IAA/Planalsucar; 1983. p. 229-264.

[28] Magalhães V. R., Influência de Doses de Vinhaça nas Características Agronômicas de Variedades de Cana-de-Açúcar, Cana-Planta e Atributos Químicos do Solo [thesis]. Universidade Estadual de Montes Claros-Unimontes; 2010.

[29] Sousa R. M. O. F., Amaral C., Fernandes J. M. C., Fraga I., Semitela S., Braga F., Sampaio A. Hazardous impact of vinasse from distilled winemaking by-products in terrestrial plants and aquatic organisms. Ecotoxicology and Environmental Safety. 2019; 183, 109493. DOI: 10.1016/j.ecoenv.2019.109493

[30] CARRILHO, E. N. V. M., LABUTO, G., KAMOGAWA, M.Y. Destination of vinasse, a residue from alcohol industry: Resource recovery and prevention of pollution. Environmental Materials and Waste: Resource Recovery and Pollution Prevention. Pp 21-43, 2016

[31] Christofoletti C. A., Escher J. P., Correia J. E., Marinho J. F. U., Fontanetti, C. S. Sugarcane vinasse: Environmental implications of its use. Waste Management. 2013; 33: 2752-2761.

[32] Marchaim, U., 1991, Biogas Processes for Sutainable Development. http://www.citeseerx,ist.tsu.edu/viewdoc/download?doi. Accessed 17 Juny 2011

[33] Van Loo S., Koppejan J. The handbook of biomass combustion and co-firing. London: Earthscan; 2008. 442 p.

[34] ASTM D2015-00, Standard Test Method for Gross Calorific Value of Coal and Coke by the Adiabatic Bomb Calorimeter (Withdrawn 2000), ASTM International, West Conshohocken, PA, 2000, www.astm.org.

[35] Gaur S., Reed T. B. An atlas of thermal data for biomass and other fuels. National Renewable Energy Lab., Golden, CO (United States); 1995.

[36] Virmond E., De Sena R.F., Albrecht, W., Althoff, C.A., Moreira R.F.P.M., José H.J. Characterization of agroindustrial solid residues as biofuels and potential application in thermochemical processes. Waste Management. 2012; 32:1952-1961.

[37] Bragato M., Joshi K., Carlson J.B., Tenório J.A.S, Levendis Y.A. Combustion of coal, bagasse and blends thereof. Part I: Emissions from batch combustion of fixed beds of fuels. Fuel. 2012; 96:43-50.

[38] Nsaful F., Görgens J.F., Knoetze J.H. Comparison of combustion and pyrolysis for energy generation in a sugarcane mill. Energy Conversion and Management. 2013; 74: 524-534.

[39] Titiloye J.O., Abu BakarM.S.,OdetoyeT.E.Thermochemical characterization of agricultural wastes from West Africa. Industrial Crops and Products. 2013; 47:199-203.

[40] Protásio T.P., Bufalino L., Tonoli G.H.D., Guimarães M., Trugilho P.F., Mendes L.M. Brazilian lignocellulosic for bioenergy production: characterization and comparison with fossil fuels. BioResources. 2013; 8:1166-1185.

[41] GEORGES, F. Caracterização da palha da cana-de-açúcar do Rio Grande do Sul e seus produtos de pirólise. 2011. 102 p. Dissertação (Mestrado em Ciência dos Materiais) - Pós- Graduação em Ciência dos Materiais, Universidade Federal do Rio Grande do Sul, PortoAlegre, 2011.

[42] Sosa-Arnao J. H. Caldeiras Aquatubulares de Bagaço – Estudo do Sistema de Recuperação de Energia. [thesis] Universidade Estadual de Campinas; 2007.

[43] Camargo J. M., Gallego-Rios J. M., Neto A. M. P., Antonio G. C., Modesto M., Toneli, J. T. Characterization of sugarcane straw and bagasse from dry cleaning system of sugarcane for cogeneration system. Renewable Energy. 2020; 158:

500-508. https://doi.org/10.1016/j.renene.2020.05.107

[44] Pelaez Samaniego M.R. Uso de biocombustível da pirólise rápida da palha de cana em um motor de ciclo Otto. [thesis].Faculdade de Engenharia Mecânica: Universidade Estadual de Campinas; 2007.

[45] De Ramos L. E., Trugilho P. F., Rezende R. N., de Assis C. O., Baliza A. E. R. Produção e avaliação de briquetes de resíduos lignocelulósicos. Pesquisa Florestal Brasileira, 2011; 31(66): 103.

[46] Pires A. A, F. Lignocellulosic residues for solid biofuel: characterization and applied thermogravimetry [thesis]. Universidade Federal de São Carlos; 2013.

[47] Perera J. G. Concentración y combustión de vinazas. Ministerio de Gobierno y Justicia de la Provincia de Tucumán, Secretaría de Estado de Gobierno y Justicia, Subsecretaria de Asuntos Técnicos; 2009.

[48] Cortez L., Rossell C. E. V., Jordan R. A., Leal M. R. L. V., Lora E. E. S. R&D needs in the industrial production of vinasse. Sugarcane bioethanol: R&D for Productivity and Sustainability. Sao Paulo, Brazil: Blucher; 619-636.

[49] M. Akram, C.K Tan, R. Garwood, S.M. Thai. Vinasse – A potential biofuel – Cofiring with coal in a fluidised bed combustor. Fuel. 2015; 158: 1006-1015.

[50] Dirbeba M.J., Brink A., DeMartini N., Zevenhoven M., Hupa M. Potential for thermochemical conversion of biomass residues from the integrated sugar-ethanol process—Fate of ash and ash-forming elements. Bioresourse Technology. 2017;234; 188-197.

[51] Gallego-Ríos J.M., Efeito das condições de processamento nas propriedades de briquetes produzidos com resíduos da produção de bioetanol: caracterização e análise do comportamento termogravimétrico [thesis]. Federal University of ABC; 2017.

[52] Palacios-Bereche M. C.; Palacios-Bereche R.; Nebra S. A. Comparison through energy, exergy and economic analyses of two alternatives for the energy exploitation of vinasse. Energy. 2020; 117231.

[53] P.B. Gangavati, M.J. Safi, A. Singh, B. Prasad, I.M. Mishra, Pyrolysis and thermal oxidation kinetics of sugar mill press mud, Thermochimica Acta. 2005; 428: 63-70. ISSN 0040-6031, https://doi.org/10.1016/j.tca.2004.09.026.

[54] Gupta N., Tripathi S., Balomajumder C. Characterization of pressmud: A sugar industry waste. Fuel. 2011; 90: 389-394. https://doi.org/10.1016/j.fuel.2010.08.021.

[55] Suresh B.N., B.S.V. Reddy. Dried Sugarcane Press Residue as a Potential Feed Ingredient Source of Nutrients for Poultry. Asian-Aust. Journal of Animal Science. 2011; 24(11): 1595-1600.

[56] Devia-OrjuelaJ.S.,Alvarez-PuglieseC. E., Donneys-Victoria D., Marriaga Cabrales N., Barba Ho L. E., Brém B., Lang D. Evaluation of Press Mud, Vinasse Powder and Extraction Sludge with Ethanol in a Pyrolysis Process. Energies. 2019; 12(21): 4145.

[57] Solomon S. Sugarcane By-Products Based Industries in India. Sugar Tech. 2011; 13: 408-416.

[58] Saidur R., Abdelaziz E.A., Dermibas A., Hossain M.S., Mekhilef S. A review on biomass as a fuel for boilers. Renewable and Sustainable Energy Reviews. 2011; 15:2262-2280.

[59] Manyà J.J., Arauzo J. An alternative kinetic approach to describe the isothermal pyrolysis of micro-particles

of sugar cane bagasse. Chemical Engineering Journal. 2008; 139:549-561.

[60] Varma, A. K., & Mondal, P. Physicochemical Characterization and Pyrolysis Kinetic Study of Sugarcane Bagasse Using Thermogravimetric Analysis. Journal of Energy Resources Technology. 2016, 138(5). DOI:10.1115/1.4032729

[61] Manatura, K. Inert torrefaction of sugarcane bagasse to improve its fuel properties. Case Studies in Thermal Engineering. 2020, 19 (100623). DOI:10.1016/j.csite.2020.100623

[62] Varma, A. K., & Mondal, P. Pyrolysis of sugarcane bagasse in semi batch reactor: Effects of process parameters on product yields and characterization of products. Industrial Crops and Products. 2017, 95: 704-717. DOI:10.1016/j.indcrop.2016.11.039

[63] Munir, S., Daood, S. S., Nimmo, W., Cunliffe, A. M., & Gibbs, B. M. Thermal analysis and devolatilization kinetics of cotton stalk, sugar cane bagasse and shea meal under nitrogen and air atmospheres. Bioresource Technology. 2009, 100(3): 1413-1418. DOI:10.1016/j.biortech.2008.07.065

[64] Cortez L. A. B., Brossard Pérez L. E. Experiences on vinasse disposal: Part III: combustion of vinasse-# 6 fuel oil emulsions. Brazilian Journal of Chemical Engineering. 1997; 14.

[65] Dasgupta A. Anaerobic digestion of solid wastes of sugar cane industry [thesis]. University of Miami; 1983.

[66] Kuruti K., Rao A. G., Gandu B., Kiran G., Mohammad S., Sailaja S., & Swamy Y. V. Generation of bioethanol and VFA through anaerobic acidogenic fermentation route with press mud obtained from sugar mill as a feedstock. Bioresource technology. 2015; 192: 646-653.

[67] Abera A. A., Duraisamy R. D., Seda T. B. Characterization of Sugar Industry Waste (Filter Cake) and Agro-waste Crop Residue as Potential Source of Livestock Feed Raw Materials; 2020.

[68] Purnomo C.W.; Salim C.; Hinode H. Preparation and characterization of activated carbon from bagasse fly ash. J Anal. Appl. Pyrolysis. 2011; 91: 257-262.

[69] Pinto C. P. Tecnologia da digestão anaeróbia da Vinhaça e desenvolvimento sustentável. Unicamp; 1999.

[70] Parikh, J., Channiwala, S. A., & Ghosal, G. K. A correlation for calculating elemental composition from proximate analysis of biomass materials. Fuel. 2007, 86(12-13): 1710-1719. DOI:10.1016/j.fuel.2006.12.029

[71] Adeniyi, A. G., Ighalo, J. O., & Abdulsalam, A. Modeling of integrated processes for the recovery of the energetic content of sugar cane bagasse. Biofuels, Bioproducts and Biorefining. 2019. DOI:10.1002/bbb.1998

[72] Lu, J.-J., & Chen, W.-H. Investigation on the ignition and burnout temperatures of bamboo and sugarcane bagasse by thermogravimetric analysis. Applied Energy. 2015, 160: 49-57. DOI:10.1016/j.apenergy.2015.09.026

[73] Decloux M.; Bories A. Stillage treatment in the French alcohol fermentation industry. International Sugar Journal. 2002.

[74] Pitarelo A.P. Avaliação da susceptibilidade do bagaço e da palha de cana-de-açúcar à bioconversão via pré-tratamento a vapor e hidrólise enzimática [thesis]. Pós-graduação em Química, Universidade Federal do Paraná, Curitiba; 2007.

[75] Klass D.L. Biomass for renewable energy, fuels and chemicals. San Diego: Academic Press, 1998. 651 p.

[76] ASTM E872-82(2013), Standard Test Method for Volatile Matter in the Analysis of Particulate Wood Fuels, ASTM International, West Conshohocken, PA, 2013, www.astm.org.

[77] Dos Santos J. R.S. Estudo da biomassa torrada de resíduos florestais de eucalipto e bagaço de cana-de-açúcar para fins energéticos. [thesis]. Escola Superior de Agricultura "Luiz de Queiroz" – ESALQ/USP; 2012.

[78] González L. M. L., Reyes I. P., Romero O. R. Anaerobic co-digestion of sugarcane press mud with vinasse on methane yield. Waste management. 2017; 68: 139-145. https://doi.org/10.1016/j.wasman.2017.07.016

[79] Saleh-e-In M. M., Yeasmin S., Paul B. K., Ahsan M., Rahman M. Z., Roy S. K. Chemical Studies on Press Mud: A Sugar Industries Waste in Bangladesh. Sugar Tech. 2012; 14(2): 109-118. https://doi.org/10.1007/s12355-012-0139-z

[80] Carvalho D. J. Estudo da viabilidade técnica de sistemas de cogeração operando exclusivamente com palha de cana. [thesis]. Faculdade de Engenharia Mecânica: Universidade Estadual de Campinas, Campinas; 2011.

Chapter 5

Sugarcane as Future Bioenergy Crop: Potential Genetic and Genomic Approaches

Muhammad Sarwar Khan, Ghulam Mustafa,
Faiz Ahmad Joyia and Safdar Ali Mirza

Abstract

Biofuels are gaining increased scientific as well as public attention to fulfill future energy demands and can be the only potential candidates to safeguard and strengthen energy security by reducing the world's reliance on exhausting fossil energy sources. Sugarcane is an important C_4 crop with great potential to contribute to global biofuel production as sugarcane juice can be easily fermented to produce ethanol. The success of bioethanol production from sugarcane in Brazil has widened the scope of the technology and has led to increased demand of purpose-grown sugarcane for biofuel production. Scientific interventions have not only helped to improve the cane crop but industrial procedures have also been upgraded resulting in improved production of bioethanol. Likewise, advancements in omics have led to high hopes for the development of energy cane. This chapter highlights the advancements as well as potential and challenges in the production of sugarcane biofuel, focusing on genetic and genomic interventions improving the crop as energy-cane. Further, controversies in the production and usage of biofuel derived from sugarcane have also been discussed.

Keywords: Biofuel, Sugarcane, Genetic and genomic approaches, Future energy-cane

1. Introduction

Increasing energy demands for erratically increasing population, urbanization, industrialization and environmental concerns related to fossil fuels have inclined researchers to explore alternative resources of energy. Compared with the current global energy consumption of 400 exajoule, an increase of 200 exajoule is expected by 2025 [1]. Biofuels have gained much importance due to the depleting fossil fuel resources and the over-accumulation of CO_2 and other greenhouse gases in the environment. Biofuels can play a part to achieve targets to replace fossil fuels to reduce carbon dioxide released into the atmosphere and to attain environmental and economic sustainability. Though bioenergy is already contributing more than 10% of global energy supplies huge potential is there to uplift its contribution. Lignocelluloses are accounted for more than 20 billion metric tons of biofuels worldwide. Owing to the outstanding features of these biofuels i.e. enhanced octane number: MON (motor octane number) and RON (research octane number), they are the most desired source of biofuel. Further, they would prove an

IntechOpen

environmentally friendly fuel source having the ability to generate less black smoke with fewer hydrocarbon emissions and NOx [2]. Hence, fuels of the renewable kind (bioethanol and biodiesel) are subjects of increased attention in this context [3]. For the development of efficient and viable alternative fuel having the ability to provide environmental safety and energy gain, the fundamental procedures need to be improved [4]. Normally, biofuels are produced either by biochemical or thermochemical strategies. Production by the biochemical method includes biomass retreatment, biomass handling, fermentation, and hydrolysis. The thermochemical processes can convert non-food and food biomass to fuel through gasification and pyrolysis [5]. A combination of both of these processes (biochemical or thermochemical) has been worked out by various research groups and has proved to be effective for the economic production of biofuel [6].

Sugarcane, being a C_4 plant is photosynthetically more competent to produce higher content of dry mass. Commercial production of sugar from sugarcane was initiated in India and China almost 2500 years back whereas it was domesticated in Western Europe during the 18th century [7]. Various Saccharum species prevail all around the world. These include *Saccharum officinarum, Saccharum edule, Saccharum barberi, Saccharum robustum, Saccharum spontaneum, and Saccharum sinense*. Among these *Saccharum spontaneum* and *Saccharum robustum* are wild species; *Saccharum officinarum*, *Saccharum barberi*, and *Saccharum sinense* are early cultivars whereas *Saccharum edule* is a marginal specialty cultivar. Anyhow, *Saccharum officinarum* is the most widely cultivated species because of its higher sucrose content and wide-spread adaptability. All the *Saccharum* genotypes are polyploid with variable ploidy level (5× to 16×) and chromosome number ranging from 80 to 130. Hence, they have the most complex genome among the plants. Researchers have been striving to develop hybrids having the ability to produce more dry mass with more sugar content. Certain hybrids have been found to have 15–25% chromosomes from *Saccharum spontaneum*, 60–70% from *Saccharum officinarum,* and 5–10% of recombinants hybrids of homologous chromosomes from both species. These hybrids are expected as better performers as compared with existing germplasm [8]. Since sugarcane is an outstanding source of biofuel production as compared with other crop plants and competitors. This manuscript highlights the significance of biofuel usage in the current scenario.

2. Potential candidate crops for biofuel production

Production of different versions of biofuels *i.e.* butanol, methanol, ethanol, isoprene, vegetable oil, hydrogen, biodiesel, jet fuel, and gasoline various types of substrates have been tested [9, 10], using renewable biomass sources including lignocellulosic sources. In the current scenario, ethanol has drawn the most attention owing to its suitability as biodiesel [11]. Various plant sources have been tested to assess their efficiency for bioethanol production and its implication as biofuel. In addition to sugarcane and maize, different plant species have shown the potential to be a valuable source for the production of biofuel (**Table 1**). Quinn et al. [13] reported the use of 49 potential plant species as feedstock for biofuel production and most of them can be used to produce ethanol. Johnsongrass, *Erianthus*, switchgrass, napiergrass, and sorghum are also valuable candidate crops for bioethanol production and can produce high biomass with fewer inputs [14]. Further, corn fiber is also a valuable source for the production of biofuel.

Perennial grasses like *Halopyrum mucronatum, Desmostachya bipinnata*, *Phragmites karka*, *Typha domingensis* and *Panicum turgidum* grows in saline coastal areas of Pakistan are good candidate plants for bioethanol production. These

Feedstock	Conditions	Biofuel production (l/ha)	References
Corn	Hydrolysis/fermentation	3,800	[12]
Sugarcane	Fermentation	7,200	
Sugar beet	Hydrolysis/fermentation	7,900	
Wheat	Hydrolysis/fermentation	1700	
Cassava	Hydrolysis/fermentation	137	

Table 1.
Conditions and comparative efficiency of different feedstocks for biofuel production.

halophytes contain a good amount of cellulose, hemicellulose, and lignin (26–37%, 24–38%, and < 10% respectively) and have a better growth rate [15]. Miscanthus a potential candidate crop for the production of bioethanol has gathered much attention because of its elemental composition, lignin and polysaccharide content, and final biomass yield. The most important property of the miscanthus is the production of the desired chemical component by thermochemical conversion because of the low ash and moisture content. These properties explored the ability and potential of this plant as a good feedstock for ethanol production in the future. Hence, the production of ethanol from edible sources faces criticism and is not economical which demands exploring non-edible plant species having the ability to grow on the marginal soils thus not interfering with the cultivation of food crops. This necessitates the utilization of saline soils to produce non-food lignocellulosic biomass which is a valuable source of bioethanol without competing for human food production [16, 17].

Another biofuel type is biodiesel that can be mixed with fossil fuels or is used directly in the engines with certain modifications [18]. The maximum oil content was determined in canola crops. Similarly, flax and camelina also appeared to be a promising source of biofuel (biodiesel) in terms of alternate energy crops. In terms of land use and competition with food crops, Camelina sativa proved to be more promising. This crop has fewer problems in comparison with Glycine max and canola and has high land-use efficiency. Further, it can be successfully grown in rotation with wheat or other winter cereals. LCA (life cycle analysis) of *C. sativa* proved that it can reduce CO_2 emission and can provide biofuel resulting in reduced consumption of fossil fuel [19]. Another valuable aspect of bioethanol is that it is an environmentally friendly fuel source compared with fossil fuels. Hence, the emis sion of greenhouse gases could be minimized by using it as a fuel source in various daily life necessities including cooking, heating, water pumping, and generation of electricity [20]. Biodiesel mixed with fossil fuel can be used in diesel engines and it does not require any kind of change in chemistry. The biodiesel produced from *B. carinata* showed prominent results and proved to be the potential crop for the production of biofuels especially in the areas where other crops are unable to provide good yield due to the adverse and variable climatic conditions [21]. *Jatropha curcas* L. also can be grown on marginal lands and can yield an oil that has been proposed to be good in performance as a biofuel. Biodiesel produced from this crop proved to be eco-friendly, biodegradable, and less toxic properties as compared with fossil fuels. The literature review highlights its contribution to therapeutic and medicinal properties as well, also, to be the source of biofuel [22]. Two wild species of perennial trees grown in the amazon were analyzed and their seed oil was examined *i.e. Carapa guianensis* and *Terminalia catappa*. Oils of the above-mentioned species were used to convert into biofuels, the resultant produced biodiesel proved to be good and comparable with the fossil fuels in terms of chemical properties and the biodiesels were acceptable for use. This showed that these species have the potential to provide oil for biodiesel production [23]. *Calophyllum inophyllum* can be the best alternative crop for the provision of oil feedstock for the production of biodiesel. Another advantage of this plant was that it can be grown in coastal regions so does not demand any sort of sacrifice in the form of food crops. The physical and chemical properties concluded that this tree has the potential to be a sustainable source of feedstock for biodiesel production. Hence, Calophyllum oil can replace palm oil to produce biodiesel. Moreover, owing to the status of a non-conventional crop it requires further research to maximize benefits from it [24]. Switchgrass is another potential crop for the production of biofuel and has gathered much interest due to its better adaptability and good performance in the field. Anyhow it needs proper attention for the economic production of biofuel from this grass [25].

3. Different generations of biofuel production (biofuels: from ethanol to biodiesel)

Biofuels are classified into different generations or groups based on their method of production and raw material used (**Figure 1**). In first-generation biofuels, cultivated crops providing polysaccharides and starch are the main raw material used in the production [26] but this generation is not much appreciated due to various reasons such as the increased amount of inputs use, cultivated crops used for biofuels are decreasing the availability of the food and cultivable land for food production. The first-generation biofuel is produced from sugars or starch. Globally, sugarcane contributes 21 million m^3 of ethanol whereas 60 million m^3 comes from corn and grains. The important step is liquefication of sugar residues followed by hydrolysis that release sugar-monomers which are converted into CO_2 and ethanol by yeast fermentation. Ethanol yield can be increased by augmenting the sugar contents of sugarcane. It is very difficult to enhance sugar production as sugarcane owns one of

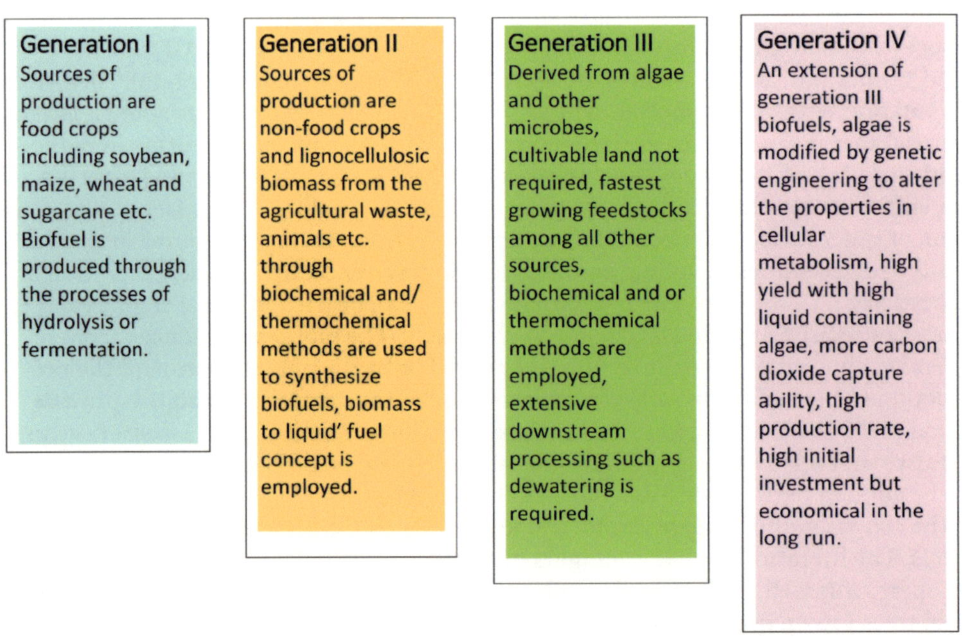

Figure 1.
Comparison among different generations of biofuel production based on their method of production and raw material used.

the most complex genomes with extremely complicated genetic networks and pathways. Moreover, modification in one process can decline the expression profiles of the other desired traits resulting in an unsolicited tradeoff. First-generation biofuels are criticized for food security as there are increasing concerns that the diversion of sugarcane towards ethanol production will reduce sugar availability, which could cause a consequent rise in prices. So, second-generation biofuels are now being used to generate energy so that co-generation would help to minimize competition with sucrose supplies. Second-generation biofuels are produced from different lignocellulose (substrate). Sugarcane bagasse and leaves are used as a by-product to produce ethanol. Various studies showed that assimilating either biochemical or thermochemical routes significantly enhances ethanol production, as compared with first-generation ethanol production. A Norwegian company (Borregaard) is the largest producer of second-generation bioethanol all over the World, with an annual production of 20,000 m^3. Milling, pretreatment (thermophysical), fermentation, distillation, and processing are the fundamental steps to produce bioethanol from lignocelluloses hence, needs due consideration to improve its production [27]. In comparison to the first generation, second-generation biofuels are cost-effective as they utilize mostly those parts of the plants that are not used as food or are considered waste. Such as crop waste or municipal wastes [28] but still there is a need to produce these biofuels at a consumer's acceptable price. Many researchers are aiming to increase the production efficiency of this type of biofuels produced but chemical or enzymatic methods are found to be more efficient. In third-generation biofuels algae is the main substrate while in fourth-generation biofuels microorganisms and genetically engineered crops and algae are used as raw material. Biofuels produced from algae are less stable and less cost-efficient. While fourth-generation biofuels are under development [29].

4. Sugarcane a bioenergy source: an overview

Energy cane, unlike conventional sugarcane, contains more fiber content than sucrose. Alexander [30] coined the term "Energy Cane Management" and proposed that sugarcane being a high biomass producer is a potential candidate for the production of bioethanol. The high biomass production is a valuable biological factor of sugarcane that contributes to the high positive LCEB (life cycle energy balance) of bioethanol produced from it with a positive balance of GHGE (greenhouse gas emission). Sugarcane biofuel generates low net greenhouse gases (GHG) and hence reduced adverse environmental impact in terms of pollution indicators. In Brazil, net greenhouse gases emission was estimated to be reduced by 25.8 million tons CO_2 equivalent in 2007. This was only because of the replacement of non-renewable energy sources with renewable energy sources [31].

Sugarcane is the most economical source of bioethanol with 9.8% fermentable sugars in its juice whereas sweet sorghum has 11.8% fermentable sugars. Sugarcane bagasse contains 22% lignin, 24% hemicellulose, and 43% cellulose whereas sweet sorghum has 21% lignin, 27% hemicellulose, and 45% cellulose with theoretical ethanol production of 12,938 and 5,804 kg per ha respectively. Though sugarcane has higher productivity as compared with its counterparts (sweet sorghum, sugar beet) yet further improvement in its fiber and sugar contents is desired to transform it into energy cane [32].

The sucrose contents present in sugarcane is about 50% of the culm dry mass whereas fiber contents are about 14%. Besides, the following features make sugarcane an ideal choice as an energy crop: drought tolerance, cold tolerance, pest and disease resistance, less flower production, erect growth habit, ratooning ability, and fast early growth [33]. Brazil is a pioneer country in launching sugarcane as an energy crop and produced about 23.4 billion liters of bioethanol in 2014 and fulfilled 15% of the country's electricity needs from sugarcane. Brazil is expecting to meet 30% (equivalent to hydropower) of energy needs from sugarcane during the current year. Thus biofuels are outcompeting the fuel market owing to their potential to fulfill future energy needs with the ability to contribute towards a safer climate. It is mandatory now in European Union and the United States to use biofuel as a fuel source. It was obligatory in the EU, to use at least 10% of the transport fuel from renewable sources by 2020. They have planned to produce 36 billion gallons of biofuel by 2022 compared with 4.7 billion gallons in 2007. Though the dominant source of production will be non-corn starch feedstock [34]. Considering this scenario, bioenergy, and energy cane seem to be the most desirable source of bioethanol all over the world.

5. Challenges in sugarcane biofuel production

Since the last two decades, biofuels have proved their worth by reducing greenhouse gases and contributing to energy self-sufficiency along with increased agricultural yield. Developed countries have enthusiastically introduced subsidies for rapid adoption of this technology but the net gains have been unsatisfactory because of rising prices of fossil fuels, high agricultural inputs, processing, transport, etc., and food crisis lead to controversy about biofuels. Researchers are of the view that competition for the production of biofuel will not only result in competition with food crops but will also lead to additional pressure for land, water, fertilizer, and other natural resources to produce feedstock. This demands reclamation of the salt-affected/waterlogged soils for the cultivation of energy crops. This will make possible the cultivation of uncultivable land resulting in enhanced production of feedstock. Further, the cultivation of neglected non-food plant species will result in their conservation thus securing biodiversity, etc. This also demands the development of developing new accession of different plant species, suitable for energy production having wide-spread adaptability to produce more biomass with least nutrients [35].

Fermentation is the most established technology for the production of biofuel from sugarcane and other feed/food crops, all over the world. In our country too, ethanol is produced from sugarcane molasses, etc. The average ethanol production is one-liter ethanol from 12 to 14 kg of sugarcane [36]. Also, bagasse is a prominent source of heat and energy. Sugarcane production requires significant investment in the form of energy and inputs including farm machinery, labor, fertilizers, and pesticides. The development of input efficient varieties with better ability to produce energy and sucrose are core challenges for the transformation of conventional sugarcane into energy cane.

6. Genome engineering in sugarcane for improvement in biofuel production

Since the advent of molecular biology, researchers are striving hard to help mankind by fulfilling their needs and uplifting living standards. Engineering plant genome for tailoring targeted cellular mechanisms and to express desired recombinant protein has proved its worth and more than 190 million hectares of agricultural land is occupied by transgenic crops. The concept of producing industrial

Sr. #	Name of the gene	Putative role of the gene	Possible outcome/ role	References
1.	*Sucrose isomerase*	Production of sucrose isomer isomaltulose	Increased sugar accumulation	[32]
2.	*CslF*	catalyze MLG biosynthesis	Cell wall biosynthesis	[49]
3.	*COMT*	Lignin reduction	Biofuel production	[50]
4.	*Phenoloxidase laccase*	Laccase enzyme activation	Detoxification of sugarcane bagasse increased second-generation biofuel	[51]
5.	*PvCCR1*	Lignin biosynthesis	Increases lignocellulosic material	[52]
6.	*COMT*	Methylation of 5- hydroxconiferyldehyde	Increases lignocellulosic material	[52]
7.	*OMT*	Pathway fluxes indicate the manipulation of the expression of a gene	Lignin modification	[53]
8.	*CAD, COMT*	LIGNIN S/G ratio	Increased biofuel production	[53]
9.	*C4H*	encode monolignol biosynthesis enzymes	Lignin biosynthesis	[54]
10.	*PAL*	Encode phenylalanine ammonia-lyase	Lignin biosynthesis	[55]
11.	*HCT*	Encodes shikimate hydroxycinnamoyl transferase	Lignin biosynthesis	[56]
12.	*Sh4CL1*	lignin biosynthesis	Lignin reduction upto 16.5%	[57]
13.	*CesA, CSL*	Lignin biosynthetic pathways	Second-generation biofuel production	[57]
14.	*CCoAOMT*	Synthesis of G and S monomer of lignin	Lignin biosynthesis	[58]
15.	*F5H*	production of the lignin S monomer	Lignin biosynthesis	[59]
16.	*CAD*	monolignol biosynthetic pathway control	Lignin biosynthesis	[59]
17.	*nptII*	Control transcriptional control of ubiquitin promoter	Increases Biomass production	[60]
18.	*CesA*	cellulose biosynthesis	Increases cellulose synthesis	[61]
19.	*Csl*	Biomass synthesis	Increase biomass synthesis	[61]
20.	*SuSy*	Cellulose synthesis	Increased cellulosic contents by 2–6%	[61]
21.	*CslH*	catalyze MLG biosynthesis	Cell wall biosynthesis	[62]
22.	*Sh4CL1*	Lignin production	Increases lignin biosynthesis	[63]

Sr. #	Name of the gene	Putative role of the gene	Possible outcome/ role	References
23.	*COMT*	Lignin biosynthesis	Second generation biofuel production	[64]
24.	*CYSOLE1, WRI1*	Supresses spd1	Increased biofuel	[64]
25.	*JU A10 T*	Lc biomass	Cellulase production	[65]
26.	*E2 ADHE, EUTE*	Liginin biosynthetic pathways	Ethanol production	[66]

Table 2.
Potential candidate genes that can be engineered to enhance biofuel production.

enzymes, therapeutics, and nutraceuticals in plants further validated the worth of plants being transformed into bio-factories. Likewise, sugarcane is an ideal candidate for the expression of desired recombinant proteins and to engineer endogenous cellular mechanisms for enhanced production of sucrose and bioethanol [8].

After addressing recalcitrance in this complex grass [37–39], efforts have been made to engineer its nuclear [40] as well as plastid genome [41] for the expression of valuable proteins in this C_4 plant. The first transgenic sugarcane was developed by Bower and Birch in 1992 [42]. Thereafter, efforts were made for herbicide tolerance [43] flowering inhibition [44], disease or pest resistance [45], drought tolerance [46], and for the expression of cellulosic enzymes in its leaves [47]. Some new genes for cold and drought tolerance are being identified in other genus and species like *S. spontaneum*, *Miscanthus*, and sorghum. Integration of these genes in sugarcane will help to generate more biomass in temperate areas or under dry conditions [48, 49]. These engineered sugarcane genotypes will provide better germplasm for the development of future energy cane for biofuel production.

Manipulating growth hormones and biomass synthesis pathways (**Table 2**) may play a vital role in plant cellulose content and total biomass leading to the development of energy cane [67]. Initial hydrolysis of cell wall polysaccharides may be increased up to 46% by reducing the cross-links of the cell wall in maize [68]. Lignin contents are the main hurdle in saccharification during conversion to ethanol and it accounts for 25% of sugarcane total lignocellulosic biomass [69]. For the saccharification process, lignin contents are needed to be removed as they prevent cellulase from accessing the cellulose molecules [67]. Almost 10 different enzymes are involved in the lignin biosynthesis pathway in sugarcane making it more complicated to engineer [70]. Some genes targeting enzymes (involved in the lignin biosynthesis pathway) like COMT (caffeic acid *O*-methyltransferase) and CAD (cinnamyl alcohol dehydrogenase) can be downregulated to alter its composition for biofuel production [71]. Biomass recalcitrance can be increased not only by lignin contents but also by lignin S/G ratio [72]. RNAi (RNA interference) suppression was used to downregulate the *COMT* gene by 67–97% to reduce lignin content and lignin S/G ratio by 3.9–13.7% and 1.47 to 1.27–0.79, respectively [50]. These findings suggested that RNAi-mediated gene suppression is a promising tool for the suppression of target genes not only involved in the lignin pathway but also the cell wall biosynthesis [73].

Biofuel production cost can be decreased by *in planta* enzyme production, as it reduces the expense of enzymes and enzyme treatment. *Cellulase* has been successfully produced in plants (i.e. *Arabidopsis*, rice, and maize) without affecting their growth and other developmental pathways [67]. However, in sugarcane *in planta* expression of enzymes is at infancy owing to its complex genome and recalcitrance. Compartmentalization of the recombinant enzymes (expression in the vacuole, chloroplast, and endoplasmic reticulum may further promote this concept, as endogenous cellular mechanisms will be least affected [47]. *Agrobacterium* and biolistic meditated transformations are well-established in sugarcane so the expression of above said enzymes in leaves and other tissues is doable [74]. Three cellulolytic enzymes (cellobiohydrolase I, cellobiohydrolase II and bacterial endoglucanase were expressed in the leaves under maize PepC promoter and with various subcellular targeting signals to assess the feasibility of accumulation of these enzymes in the vacuole, endoplasmic reticulum, and chloroplast. Expression of cellobiohydrolase (CBHs) was maximum in the vacuoles whereas expression of endoglucanase was maximum in the chloroplasts. Hence, these studies proved that the sugarcane genome may be targeted for the expression of cellulolytic enzymes leading to the economical production of bioethanol [47].

7. Role of omics in the development of future energy cane

Omics is genuinely an innovative area of research in the field of genomics, transcriptomics, proteomics, metabolomics, and their applications for the improvement of sugarcane to energy cane. Understanding the genetic regulation and mechanisms involved in photosynthesis, nutrient assimilation, disease resistance, sucrose transportation. Advancement in genome mapping, DNA microarray, expression profiling, RNAi (RNA interference), and data mining tools can play a central role in the development of future energy cane. Interventions in next-generation sequencing (NGS) has not only reduced the cost of whole-genome sequencing but have made it possible to sequence complex genomes like sugarcane [75]. It is of particular importance for the aneuploid, polyploid genome of sugarcane for the identification of various alleles of the same gene. This has also helped to devise molecular markers paving the way to tackle bottlenecks in sugarcane breeding. Numerous transcript sequence clusters of sugarcane do not contain information of full-length coding sequences, NGS is expected to resolve the issue by making it possible to get information of complete gene sequences [76].

With the advent of recent techniques, conventional breeding can now be integrated with genomic tools to harvest maximum advantage of these innovations. The earlier genomic research led to the development of molecular markers, elucidated genome structure of modern genotypes, and phylogenetic relationship among the complex Saccharum species. EST-SSRs have been successfully utilized to understand genetic relationships and genetic diversity. Genome mapping research has helped to determine marker-trait associations and to validate chromosomal localization of valuable genes [77]. The development of new markers and their incorporation in genetic maps will accelerate breeding programs leading to the development of an approved version of sugarcane. Understanding complex connections among genetics, genes, proteome, and metabolome requires integrated research on omics, bioinformatics, and computational biology. A great many sugarcane genes involved in molecular mechanisms of stress (cold, drought, and salinity stress), plant growth, and development have been explored [78]. During the recent decade, transcriptomic research has led to the identification of more than 33000 genes, involved in critical biological functions in this energy crop [79].

Plant genomic databases are valuable resources to mine candidate genes for the improvement of crop plants through molecular breeding. Many databases have been developed for various plant species. These include Gramene [80] TropGENE [81], Plant GDB [82], GRASSIUS [83], Phytozome [84], MOROKOSHI [85], Plant TF database [86], PLAZA [87] and KBase [88]. SUCESTFUN is a specific platform for sugarcane/energy cane breeders (http://sucest-fun.org/) [89] and was developed keeping in view the five major objectives: gene annotation, expression profiling, genome sequencing, functional genomics, and integration of public resources. The database was initially based on 43,141 Sugarcane Assembled Sequences (SAS) from the SUCEST Project [90] followed by the generation of 17,500 ORFeome genes using RNA-seq from the hybrid and ancestral varieties of sugarcane [91]. The retrieved data was of great value for single nucleotide polymorphism analysis, protein characterization, identification of splicing variants, and evolutionary studies.

The first transcriptomic analysis was performed to investigate differences between immature and mature leaves and internodes, [92] followed by the genes involved in sucrose transportation, source-sink interactions, tissue profiling of transcriptionally active transposable elements, transcription elements, stress-responsive factors, and resistance gene analogs [93]. Numerous differentially expressed transcripts involved in photosynthesis, assimilate partitioning, cell wall synthesis, phosphate metabolism, and stress were identified through an oligonucleotide array [94]. The advent of micro-arrays appeared a great milestone in expression profiling where GeneChip^R Sugar Cane Genome Array produced by Affymetrix assaying up to 4715 non-redundant random ESTs.

The available sugarcane genomic resources can be employed to identify genes involved in sugarcane cell wall (SCW) biosynthesis. NAC and MYB transcription factors and gene regulatory network (GRN) involved in SCW biosynthesis. The genes identified through genomic and transcriptomic approaches could either be used as DNA markers or develop value-added transgenic sugarcane [95]. Proteomics approaches have explored the role of DEPs (differentially expressed proteins) in signal transduction pathways for stress tolerance by proteomic approaches i.e. 2D-DIGE (two-dimensional difference gel electrophoresis) [96] and iTRAQ (isobaric tags for relative and absolute quantitation) [97]. More recently, metabolite analysis provides a deeper understanding of the complex regulatory processes of potential metabolites including saccharides and other derivatives helping out to predict resistance mechanisms through the use of high-throughput technologies that can determine metabolic phenotypes [98, 36]. Hence, these advancements can play a crucial role in the development of future energy cane (**Figure 2**).

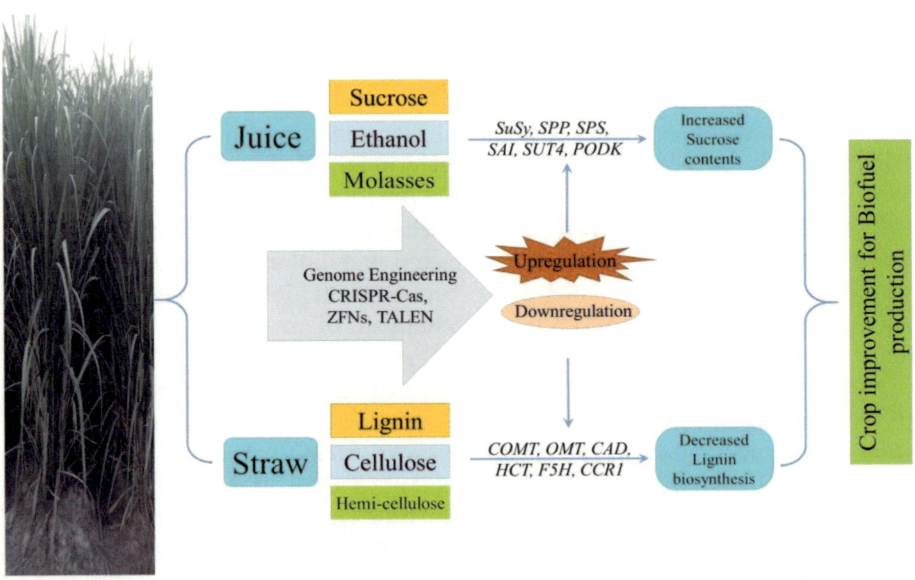

Figure 2.
Schematic sketch showing different research strategies to uplift biofuel production from sugarcane.

8. Conclusions

Fuel energy is an inevitable necessity of life and is anticipated to gain key standing in the world's economy. Exhausting reserves and the ruthless burning of fossil fuels have forced researchers to explore alternative energy resources for the biosecurity of living beings. Biofuels are potential candidates that provide an eco-friendly and sustainable energy source to meet the energy demand of the whole world. The major contributors to global biofuel are maize and sugarcane in addition to sorghum and other grasses. Research efforts are direly needed, not only to increase biomass production but also the betterment of industrial processes involved in the production of biofuel including biodiesel. Advancements in omics and other innovative disciplines have opened new horizons paving the way to develop future energy crops resulting in the replacement of fossil energy with renewable energy.

Author details

Muhammad Sarwar Khan[1], Ghulam Mustafa[1], Faiz Ahmad Joyia[1] and Safdar Ali Mirza[2]*

1 Centre of Agricultural Biochemistry and Biotechnology (CABB), University of Agriculture, Faisalabad, Pakistan

2 Government College University, Lahore, Pakistan

*Address all correspondence to: safdaralimirza@gcu.edu.pk

IntechOpen

© 2021 The Author(s). Licensee IntechOpen. This chapter is distributed under the terms of the Creative Commons Attribution License (http://creativecommons.org/licenses/by/3.0), which permits unrestricted use, distribution, and reproduction in any medium, provided the original work is properly cited.

References

[1] Ahorsu R, Medina F, Constantí M. Significance and Challenges of Biomass as a Suitable Feedstock for Bioenergy and Biochemical Production: A Review. Energies. 2018;11:3366

[2] Valderrama C, Quintero V, Kafarov V. Energy and water optimization of an integrated bioethanol production process from molasses and sugarcane bagasse: A Colombian case. Fuel 2020;260:116314.

[3] Tareen WU, Dilbar MT, Farhan M, Ali Nawaz M, Durrani AW, Memon KA, Mekhilef S, Seyedmahmoudian M, Horan B, Amir M, Aamir M. Present Status and Potential of Biomass Energy in Pakistan Based on Existing and Future Renewable Resources. Sustainability 2020;12:249

[4] Hill J, Nelson E, Tilman D, Polasky S, Tiffany D. Environmental, economic, and energetic costs and benefits of biodiesel and ethanol biofuels. PNAS 2006;103:11206-11210.

[5] Naqvi SR, Uemura Y, Yusup S, Nishiyama N. Catalytic consequences of micropore topology on biomass pyrolysis vapors over shape-selective zeolites. Energy Procedia 2017;105:557-561.

[6] Ubando AT, Chen WH, Tan RR, Naqvi SR. Optimal integration of a biomass-based polygeneration system in an iron production plant for negative carbon emissions. Int. J. Energy Res. 2019. https://doi.org/10.1002/er.4902.

[7] Grivet L, Daniels C, Glaszmann JC and D'Hont A. A review of recent molecular genetics evidence for sugarcane evolution and domestication. Ethnobotany research and application. 2004;2:9-17.

[8] Arruda P. Genetically modified sugarcane for bioenergy generation. Current opinion in biotechnology. 2012;23:315-322

[9] Chew KW, Chia SR, Show PL, Ling TC, Arya SS, Chang J-S. Food waste compost as an organic nutrient source for the cultivation of Chlorella vulgaris. Bioresource technology 2018;267:356-362

[10] Chia SR, Ong HC, Chew KW, Show PL, Phang S-M, Ling TC, Nagarajan D, Lee D-J, Chang J-S. Sustainable approaches for algae utilisation in bioenergy production. Renewable energy 2018;129:838-852.

[11] Smith EG, Janzen HH, Newlands NK. Energy balances of biodiesel production from soybean and canola in Canada. Canadian journal of plant science. 2007;87:793-801.

[12] Lin Y, Tanaka S. Ethanol fermentation from biomass resources: Current state and prospects. Applied microbiology and biotechnology. 2006;69(6):627-642.

[13] Quinn LD, Gordon DR, Glaser A, Lieurance D, Flory SL. Bioenergy Feedstocks at Low Risk for Invasion in the USA: a "White List" Approach. Bioenergy research. 2015;8:471-481

[14] Brosse N, Dufour A, Meng X, Sun Q, Ragauskas A. Miscanthus: a fast-growing crop for biofuels and chemicals production. Biofuels, Bioproducts and Biorefining. 2012;6:580-598.

[15] Abideen Z, Ansari R, Khan MA. Halophytes: Potential source of ligno-cellulosic biomass for ethanol production. Biomass and bioenergy. 2011;35:1818-1822.

[16] Malode SJ, Prabhu KK, Mascarenhas RJ, Shetti NP, Aminabhavi TM. Recent advances and viability in biofuel production. Energy

Conversion and Management: X 2021; 10:100079.

[17] Abideena Z, Hameeda A, Koyrob HW, Gula B, Ansaria R, Khan MA. Sustainable biofuel production from non-food sources – An overview. Emir. J. Food Agric. 2014;26:1057-1066.

[18] Murugesan A, Umarani C, Subramanian R, Nedunchezhian N. Bio-diesel as an alternative fuel for diesel engines—a review. Renewable and sustainable energy reviews. 2009;13:653-662.

[19] Blackshaw R, Johnson E, Gan Y, May W, McAndrew D, Barthet V, McDonald T, Wispinski D.. Alternative oilseed crops for biodiesel feedstock on the Canadian prairies. Canadian Journal of Plant Science. 2011;91:889-896.

[20] Leduc S, Natarajan K, Dotzauer E, McCallum I, Obersteiner M. Optimizing biodiesel production in India. Applied Energy 2009;86:S125-S131.

[21] Cardone M, Mazzoncini M, Menini S, Rocco V, Senatore A, Seggiani M, Vitolo S. Brassica carinata as an alternative oil crop for the production of biodiesel in Italy: agronomic evaluation, fuel production by transesterification and characterization. Biomass and Bioenergy. 2003;25:623-636.

[22] Pandey VC, Singh K, Singh JS, Kumar A, Singh B, Singh RP. Jatropha curcas: A potential biofuel plant for sustainable environmental development. Renewable and Sustainable Energy Reviews. 2012;16:2870-2883

[23] Iha OK, Alves FC, Suarez PA, Silva CR, Meneghetti MR, Meneghetti SM. Potential application of Terminalia catappa L. and Carapa guianensis oils for biofuel production: Physical-chemical properties of neat vegetable oils, their methyl-esters and bio-oils (hydrocarbons). Industrial Crops and Products. 2014;52:95-98

[24] Ong H, Mahlia T, Masjuki H, Norhasyima R. Comparison of palm oil, Jatropha curcas and Calophyllum inophyllum for biodiesel: a review. Renewable and Sustainable Energy Reviews. 2011;15:3501-3515

[25] David K, Ragauskas AJ. Switchgrass as an energy crop for biofuel production: a review of its ligno-cellulosic chemical properties. Energy & Environmental Science. 2010;3:1182-1190

[26] Alalwan HA, Alminshid AH, Aljaafari HA. Promising evolution of biofuel generations. Subject review. Renewable Energy Focus. 2019;28:127-139

[27] Paschalidou A, Tsatiris M, Kitikidou K. Energy crops for biofuel production or for food?-SWOT analysis (case study: Greece). Renewable Energy. 2016;93:636-647.

[28] Rodionova MV, Poudyal RS, Tiwari I, Voloshin RA, Zharmukhamedov SK, Nam HG, Zayadan BK, Bruce BD, Hou HJ, Allakhverdiev SI. Biofuel production: challenges and opportunities. International Journal of Hydrogen Energy. 2017;42(12):8450-8461.

[29] Saini JK, Saini R, Tewari L. Lignocellulosic agriculture wastes as biomass feedstocks for second-generation bioethanol production: concepts and recent developments. 3 Biotech 2015;5:337-353

[30] Alexander AG. Energy cane as a multiple-products alternative. In: Proceedings of the Post-Conference Seminar, Aiea Library, p. 17, Hawaii Natural Energy Institute, Aiea, Hawaii, USA, November 1984.

[31] Beig B, Riaz M, Naqvi SR, Hassan M, Zheng Z, Karimi K, Pugazhendhi A, Atabani AE, Chi NTL. Current challenges and innovative developments in pretreatment of lignocellulosic residues for biofuel production: A review. Fuel. 287; 119670.

[32] Kandel R, Yang X, Song J, Wang J. Potentials, challenges, and genetic and genomic resources for sugarcane biomass improvement. Frontier in plant science. 2018;9:151.

[33] de Souza AP, Grandis A, Leite DCC, Buckeridge MS. Sugarcane as a Bioenergy Source: History, Performance, and Perspectives for Second-Generation Bioethanol. Bioenergy research. 2014;7(1):24-35

[34] Pearson RJ, Turner JWG. Renewable Fuels: An Automotive Perspective. In: Roddy DJ. (ed.) Comprehensive Renewable Energy. Elesvier; 2012. p.305-342

[35] Keyzer MA, Merbis MD and Voortman RL. The biofuel controversy De Economist. 2008;156(4):507-527

[36] Patzek TW, Pimentel D. Thermodynamics of energy production from Biomass: Critical reviews in plant sciences. 2005;24(5):327-364.

[37] Ali A, Khan M, Sharif R, Mujtaba M, Gao SJ. Sugarcane omics: an update on the current status of research and crop improvement. Plants. 2019;8:344

[38] Mustafa G, Khan MS. Reproducible in vitro regeneration system to purify sugarcane clones. African Journal of Biotechnology. 2012;11(42):9961-9969.

[39] Mustafa, G, Khan MS. Differential role of indolebutyric acid in sugarcane root development. Sugar Tech. 2016;18:55-60.

[40] Khan MS, Ali S, Iqbal J. Developmental and photosynthetic regulation of Bacillus thuringiensis δ-endotoxin reveals that engineered sugarcane conferring resistance to 'dead heart' contains no toxins in cane juice. Molecular Biology Reports. 2011;38: 2359-2369

[41] Mustafa G, Khan MS. Transmission of engineered plastids in Sugarcane, a C_4 monocotyledonous plant, reveals that sorting of preprogrammed progenitor cells produce heteroplasmy. Plants. 2021;10;26

[42] Bower R, Birch RG. Transgenic sugarcane plants via microprojectile bombardment. The plant journal. 1992;2:409-416.

[43] Enríquez-Obregón GA, Vázquez-Padrón RI, Prieto-Samsonov DL, De La Riva GA, Selman-Housein G. Herbicide-resistant sugarcane (*Saccharum officinarum* L.) plants by Agrobacterium-mediated transformation. Planta. 1998;206:20-27

[44] Hotta CT, Lembke CG, Domingues DS, Ochoa EA, Cruz GM, Melotto-Passarin DM, Marconi TG, Santos MO, Mollinari M, Margarido GR, Crivellari AC. Van Sluys Marie-Anne, Souza GM. The biotechnology roadmap for sugarcane improvement. Tropical plant biology. 2011;3:75-87

[45] Zhang L, Xu J, Birch RG. Engineered detoxification confers resistance against a pathogenic bacterium. Nature biotechnology. 1999;17:1021-1024.

[46] Zhang, S.-Z., Yang, B.-P., Feng, C.-L., Chen, R.-K., Luo, J.-P., Cai, W.-W., et al. Expression of the *Grifola frondosa* trehalose synthase gene and improvement of drought-tolerance in sugarcane (*Saccharum officinarum* L.). Journal of integrative plant biology. 2006;48:453-459

[47] Harrison MD, Geijskes J, Coleman HD, Shand K, Kinkema M, Palupe A, Hassall R, Sainz M, Lloyd R, Miles S, Dale JL. Accumulation of recombinant cellobiohydrolase and endoglucanase in the leaves of mature transgenic sugar cane. Plant biotechnology journal. 2011;9:884-896.

[48] Mustafa G, Joyia FA, Anwar S, Parvaiz A, Khan MS. Biotechnological interventions for the improvement of sugarcane crop and sugar production. In: De Oliveira A (ed.) Sugarcane - Technology and Research. InTech-Open Science. 2018. p113-138

[49] Lam E, Shine Jr J, Da Silva J, Lawton M, Bonos S, Calvino M, Carrer H, Silva-Filho MC, Glynn N, Helsel Z, Ma J, Richard Jr E, Souza GM, Ming R. Improving sugarcane for biofuel: engineering for an even better feedstock," GCB Bioenergy. 2009;1(3):251-255

[50] Jung JH, Fouad WM, Vermerris W, Gallo M, Altpeter F. RNAi suppression of lignin biosynthesis in sugarcane reduces recalcitrance for biofuel production from lignocellulosic biomass. Plant biotechnology journal. 2012;10:1067-1076.

[51] Harish BS, Ramaiah MJ, Uppuluri KB. Bioengineering strategies on catalysis for the effective production of renewable and sustainable energy. Renewable and sustainable energy reviews. 2015;51:533-547

[52] Simmons BA, Loqué D, Ralph J. Advances in modifying lignin for enhanced biofuel production. Current opinion in plant biology. 2010;13(3):312-319.

[53] Hoang NV, Furtado A, Botha FC, Simmons BA, Henry RJ. Potential for genetic improvement of sugarcane as a source of biomass for biofuels. Frontiers in bioengineering and biotechnology. 2015;3:1-15

[54] Vaish B, Srivastava V, Singh P, Singh P, Singh R. Energy and nutrient recovery from agro-wastes: Rethinking their potential possibilities. Environmental engineering research. 2020; 25(5):623-637

[55] Santos FA, Queiróz JH, Colodette JL, Fernandes SA, Guimarães VM, Rezende ST. Potential of sugarcane straw for ethanol production. Química nova. 2012;35(5):1004-1010

[56] Yuan JS, Tiller KH, Al-Ahmad H, Stewart NR, Stewart Jr CN. Plants to power: bioenergy to fuel the future. Trends in plant science. 2008;13(8): 421-429

[57] Schmatz AA, Tyhoda L, Brienzo M. Sugarcane biomass conversion influenced by lignin. Biofuels bioproducts and biorefining. 2020;14(2):469-480

[58] Kasirajan L, Hoang NV, Furtado A, Botha FC, Henry RJ. Transcriptome analysis highlights key differentially expressed genes involved in cellulose and lignin biosynthesis of sugarcane genotypes varying in fiber content. Scientific reports. 2018;8(1):1-16

[59] Jung JH, Altpeter F. TALEN mediated targeted mutagenesis of the caffeic acid O-methyltransferase in highly polyploid sugarcane improves cell wall composition for production of bioethanol. Plant molecular biology. 2016;92(1-2):131-142

[60] Jung JH, Vermerris W, Gallo M, Fedenko JR, Erickson JE, Altpeter F. RNA interference suppression of lignin biosynthesis increases fermentable sugar yields for biofuel production from field-grown sugarcane. Plant biotechnology journal. 2013;11(6):709-716

[61] Xie G, Peng L. Genetic engineering of energy crops: a strategy for biofuel production in china free access. Journal

of Integrative Plant biology. 2011;53(2):143-150.

[62] Bush DR, Leach JE. Translational genomics for bioenergy production: There's room for more than one model. Plant cell. 2007;19(10):2971-2973

[63] Buckeridge MS, De Souza AP, Arundale RA, Anderson-Teixeira KJ, DeLucia E. Ethanol from sugarcane in Brazil: A 'midway' strategy for increasing ethanol production while maximizing environmental benefits. GCB Bioenergy, 2012;4(2):119-126

[64] Parajuli S, Kannan B, Karan R, Sanahuja G, Liu H, Garcia-Ruiz E, Kumar D, Singh V, Zhao H, Long S, Shanklin J. Towards oilcane: Engineering hyperaccumulation of triacylglycerol into sugarcane stems. GCB Bioenergy. 2020;12(7):476-490.

[65] Singh A, Patel AK, Adsul M, Singhania RR. Genetic modification: A tool for enhancing cellulase secretion. Biofuel research journal. 2017;4(2): 600-610

[66] Dhaliwal SS, Oberoi HS, Sandhu SK, Nanda D, Kumar D, Uppal SK. Enhanced ethanol production from sugarcane juice by galactose adaptation of a newly isolated thermotolerant strain of Pichia kudriavzevii. Bioresource technology. 2011;102(10):5968-5975

[67] Sticklen M. Plant genetic engineering to improve biomass characteristics for biofuels. Current opinion in biotechnology. 2006;17:315-319.

[68] Grabber JH. How do lignin composition, structure, and cross-linking affect degradability? A review of cell wall model studies. Crop science. 2005;45:820

[69] Canilha L, Rodrigues R, Antunes FAF, Chandel AK, Milessi T, Felipe MDGA, et al. (2013). Bioconversion of hemicellulose from sugarcane biomass into sustainable products in Sustainable Degradation of Lignocellulosic Biomass-Techniques, Applications and Commercialization, eds A. K. Chandel and S. S. da Silva (Rijeka: InTech), 15-45. doi:10.5772/53832

[70] Whetten, R., and Ron, S. Lignin biosynthesis. Plant cell. 1995;7: 1001-1013.

[71] Furtado A, Lupoi JS, Hoang NV, Healey A, Singh S, Simmons BA, Henry RJ. Modifying plants for biofuel and biomaterial production. Plant biotechnology journal. 2014;12:1246-1258

[72] Li X, Ximenes E, Kim Y, Slininger M, Meilan R, Ladisch M, Chapple C. Lignin monomer composition affects Arabidopsis cell-wall degradability after liquid hot water pretreatment. Biotechnology for biofuels. 2010;3:27.

[73] Bottcher A, Cesarino I, dos Santos AB, Vicentini R, Mayer JL, Vanholme R, Morreel K, Goeminne G, Moura JC, Nobile PM, Carmello-Guerreiro SM. Lignification in sugarcane: biochemical characterization, gene discovery, and expression analysis in two genotypes contrasting for lignin content. Plant physiology. 2013;163:1539-1557.

[74] Taylor II LE, Dai Z, Decker SR, Brunecky R, Adney WS, Ding SY, Himmel ME. Heterologous expression of glycosyl hydrolases in planta: a new departure for biofuels. Trends in biotechnology. 2008;26:413-424

[75] Grativol C, Regulski M, Bertalan M, McCombie WR, da Silva FR, Zerlotini Neto A, Vicentini R, Farinelli L, Hemerly AS, Martienssen RA, Ferreira PC. Sugarcane genome sequencing by methylation filtration

provides tools for genomic research in the genus Saccharum. The Plant Journal. 2014;79;162-172

[76] Diniz AL, Sávio S. Ferreira b, Felipe ten-Caten a, Gabriel R.A. Margarido c, João M. dos Santos d, Geraldo V. de S. Barbosa d, Monalisa S. Carneiro e, Glaucia M. Souza. Genomic resources for energy cane breeding in the post genomics era. Computational and Structural Biotechnology Journal. 2019;17:1404-1414

[77] Khueychai S, Jangpromma N, Daduang S, Jaisil P, Lomthaisong K, Dhiravisit A, Klaynongsruang S. Comparative proteomic analysis of leaves, leaf sheaths, and roots of drought-contrasting sugarcane cultivars in response to drought stress. Acta physiologiae plantarum. 2015;37:88

[78] Kaur L, Dharshini S, Ram B, Appunu C. Sugarcane Genomics and Transcriptomics. In: Sugarcane Biotechnology: Challenges and Prospects; Springer: Cham, Switzerland, p.13-32.

[79] Augustine SM, Narayan JA, Syamaladevi DP, Appunu C, Chakravarthi M, Ravichandran V, Tuteja N, Subramonian N. Overexpression of EaDREB2 and pyramiding of EaDREB2 with the pea DNA helicase gene (PDH45) enhance drought and salinity tolerance in sugarcane (Saccharum spp. hybrid). Plant cell reports. 2015;34:247-263

[80] Ware D, Jaiswal P, Ni J, Pan X, Chang K, Clark K, Teytelman L, Schmidt S, Zhao W, Cartinhour S, McCouch S. Gramene: a resource for comparative grass genomics. Nucleic acids research. 2002;30:103-105.

[81] Ruiz M, Rouard M, Raboin LM, Lartaud M, Lagoda P, Courtois B. TropGENE-DB, a multi-tropical crop information system. Nucleic acids research. 2004;32:D364–D367

[82] Duvick J, Fu A, Muppirala U, Sabharwal M, Wilkerson MD, Lawrence CJ, Lushbough C, Brendel V. PlantGDB: a resource for comparative plant genomics. Nucleic acids research. 2008;36:D959–D965

[83] Yilmaz A, Nishiyama MY, Fuentes BG, Souza GM, Janies D, Gray J, Grotewold E. GRASSIUS: a platform for comparative regulatory genomics across the grasses. Plant physiology. 2009;149:171-180

[84] Goodstein DM, Shu S, Howson R, Neupane R, Hayes RD, Fazo J, Mitros T, Dirks W, Hellsten U, Putnam N, Rokhsar DS. Phytozome: a comparative platform for green plant genomics. Nucleic acids research. 2012;40:D1178–D1186.

[85] Makita Y, Shimada S, Kawashima M, Kondou-Kuriyama T, Toyoda T, Matsui M. MOROKOSHI: transcriptome database in sorghum bicolor. Plant cell physiology. 2015;56:e6.

[86] Jin J, Tian F, Yang DC, Meng YQ, Kong L, Luo J, Gao G. PlantTFDB 4.0: toward a central hub for transcription factors and regulatory interactions in plants. Nucleic acids research. 2017;45:D1040–D1045

[87] Van Bel M, Diels T, Vancaester E, Kreft L, Botzki A, Van de Peer Y, Coppens F, Vandepoele K. 4.0: an integrative resource for functional, evolutionary and comparative plant genomics. Nucleic acids research. 2018;46:D1190–D1196.

[88] Arkin AP, Cottingham RW, Henry CS, Harris NL, Stevens RL, Maslov S, Dehal P, Ware D, Perez F, Canon S, Sneddon MW. KBase: the United States Department of Energy Systems Biology Knowledgebase. National biotechnology. 2018;36:566-569

[89] Nishiyama-Jr MY, Vicente FF, Lembke CG, Sato PM, Dal-Bianco ML, Fandino RA, Hotta CT, Souza GM. The SUCEST-FUN regulatory network database: designing an energy grass. International sugar journal. 2012;114:821-826.

[90] Vettore AL, da Silva FR, Kemper EL, Souza GM, da Silva AM, Ferro MIT, Henrique- Silva F, Vettore AL. Analysis and functional annotation of an expressed sequence tag collection for tropical crop sugarcane. Genome research. 2003;13:2725-2735.

[91] Nishiyama MY, Ferreira SS, Tang P-Z, Becker S, Pörtner-Taliana A, Souza GM. Full-length enriched cDNA libraries and ORFeome analysis of sugarcane hybrid and ancestor genotypes. PLoS ONE. 2014;9

[92] Carson D, Huckett B, Botha F. Differential gene expression in sugarcane leaf and internodal tissues of varying maturity. South African Journal of Botany. 2002;68:434-442

[93] Parvaiz A, Mustafa G, Khan HMWA, Joyia FA, Niazi AK, Anwar S, Khan, MS. Field evaluation ratified by transcript and computational analyses unveils myco-protective role of SUGARWIN proteins in sugarcane. 3 Biotech. 2019;9:377.

[94] McCormick A, Cramer M, Watt D. Changes in photosynthetic rates and gene expression of leaves during a source sink perturbation in sugarcane. Annals of botany. 2008;101:89-102.

[95] Li M, Liang Z, Zeng Y, Jing Y, Wu K. Liang J. He S. Wang G. Mo Z. Tan F (2016) De novo analysis of transcriptome reveals genes associated with leaf abscission in sugarcane (*Saccharum oficinarum* L.). BMC Genom., 17, 195.

[96] Barnabas L, Ramadass A, Amalraj RS, Palaniyandi M, Rasappa V. Sugarcane proteomics: An update on current status, challenges, and future prospects. Proteomics. 2015;15:1658-1670

[97] Su Y, Xu L, Wang Z, Peng Q, Yang Y, Chen Y, Que Y. Comparative proteomics reveals that central metabolism changes are associated with resistance against Sporisorium scitamineum in sugarcane. BMC Genom. 2016;17:1-21

[98] Lee DK, Ahn S, Cho HY, Yun HY, Park JH, Lim J, Lee J, Kwon SW. Metabolic response induced by parasitic plant-fungus interactions hinder amino sugar and nucleotide sugar metabolism in the host. Scientific reports. 2016;6:1-11

Chapter 6

Potential of Bagasse as Raw Material for Lignosulfonate Surfactant

Rini Setiati, Aqlyna Fatahanissa, Shabrina Sri Riswati, Septoratno Siregar and Deana Wahyuningrum

Abstract

Anionic surfactants are generally used in surfactant injections because they are good, resistant in storage and stable. Furthermore, Commercially, anions are produced in the form of carboxylates, sulfates, sulfonates, phosphates, or phosphonates. The surfactants used in the process of implementing Enhanced Oil Recovery (EOR) are generally petroleum-based, such as Petroleum Sulfonate. Therefore, an increase in oil price, leads to an increase in the price of surfactant and the operational costs becomes relatively expensive. Lignosulfonate is a type of anionic surfactant which is made with lignin as raw material. This lignin is found in many plants, including wood stalks, plant leaves, peanut shells, corn cobs, bagasse, empty bunches of oil palm and wheat straw. Based on the results of previous studies, 25% of lignin component was discovered in bagasse. This may be a consideration that there is enough lignin in bagasse to be used as raw material in the production of lignosulfonate vegetable surfactants. Furthermore, lignin from bagasse is used because bagasse is easy to obtain, cheap and an environmental friendly vegetable waste. Currently, bagasse is only used as fuel in steam boilers and papermaking, cement and brick reinforcement, a source of animal feed, bioethanol, activated charcoal as adsorbent and compost fertilizer. This is a consideration to optimize the use of bagasse to become lignosulfonate as an alternative for surfactants in the petroleum sector. The purpose of this study is to show that lignin from bagasse has the potential of becoming a lignosulfonate surfactant. There are several studies that have processed bagasse into sodium lignosulfonate. The component test on the results showed that the surfactant component of sodium lignosulfonate from bagasse was almost the same as the commercial standard lignosulfonate component. Furthermore, the results of the HLB (Hydrophilic–Lipophilic Balance) value test show that the sodium lignosulfonate surfactant from bagasse can function as an emulsion form which is a required parameter for the surfactant injection mechanism. Based on the discussion of the study results, bagasse has the potential as a raw material to be processed into lignosulfonates.

Keywords: bagasse, lignin, lignosulfonate, surfactant

1. Introduction

Indonesia is an agricultural country centered on the equatorial landscape where the longest, longest and most photosynthetic process occurs throughout the year. The initial product of photosynthesis is glucose which is synthesized from carbon dioxide (CO_2) and water with the help of sunlight in chlorophyll. High rainfall guarantees water availability and the progress of the oxidation process. The process of decomposition in tropical climates occurs at a fast rate so that there is enough CO_2 in the air. With radiation for about 10 hours a day, Indonesia is one of the most productive regions in the world. As a large archipelago, Indonesia also has vast fertile land for the cultivation of sugar-producing crops. Sugar-producing plants that grow and develop well in the tropics include sugar cane, palms (palm, coconut), and beet plants [1].

Sugarcane is a plant used as a raw material in the production of sugar. It has high sweet content on the stem and holds many benefits behind its distinctive sweet taste which are not only in terms of health but from various terms, such as industry, household consumption, agriculture and livestock. This plant consists of a grass that has many types and varieties, ranging from yellow, red, e.t.c. [2]. Sugarcane harvest is not influenced by the season where the harvest is still satisfactory, which is during the transition season [3]. The stalks of sugarcane harvested from plantations are trucked to the factory to be processed into sugar [4].

Sugarcane is a type of plant grown only in areas with tropical climates. In Indonesia, sugarcane plantations have an area around 400 to 500 hectares thousand hectares spread across Medan, Lampung, Solo, Tegal and Mojokerto [5]. Sugarcane as a raw material for the sugar industry is one of the plantation commodities that has a strategic role in the economy in Indonesia. With an area of approximately 415.66 thousand hectares in 2018, the sugar cane industry is a source of income for thousands of sugarcane farmers and workers in the sugar industry [6]. **Figure 1** below shows the conditions of the sugarcane plantation until the sugarcane is ready for harvest. Sugarcane harvesting is carried out by cutting the sugarcane stalks by workers and then preparing the clean sugarcane stalks to be processed into sugar or sugarcane juice with a sweet taste [7].

Figure 1.
Sugarcane from plants to consumers.

The main product of sugarcane is its extract which is used as the main ingredient in producing sugar. On a large scale, majority of sugar cane is used in the production of white and brown sugar. Sugarcane that enters the factory must be sugarcane that is ripe or has high brix and pol and is clean from all types of impurities (roots, shoots, dry leaves) [8–10]. Manual fellers, the results are better than sugarcane harvesting machines. Logging covers all parts of the sugarcane. The shoots and leaves are discarded and only the sugarcane stalks are used because what contains sucrose is sugarcane stalks. The cleanliness of sugarcane from manure depends on the skills of felling personnel and the application of felling SOPs in the field. In addition, milled sugarcane must be fresh, fresh sugarcane is sugarcane that is milled for no more than 48 hours [11, 12]. The distribution starts from sugarcane harvested from plantations [13, 14]. Sugarcane loaded onto trucks by humans and machines [15, 16]. And then transported to the sugar factory as shown in the following **Figures 2** and **3** below.

In **Figure 2**, you can see the process of transporting sugarcane, starting from cutting down sugarcane stalks by workers, collecting sugarcane stalks by workers or using grab folders, to transporting them to trucks either by workers or using grab folders. After the sugarcane is loaded into the truck, it is taken to the sugar factory to be processed. The sugarcane that has arrived at the sugar factory will be poured into the mill as shown in **Figure 3** below. The loading of sugarcane into the mill has used a system in such a way that the truck has been facilitated so that it can directly load the sugarcane mill [17, 18].

In **Figure 3**, there are no human workers who move sugarcane stalks to the sugar mill. Inside the factory, the sugarcane stalks harvested from plantations are processed into sugar by undergoing a five-time grinding process, after which it is

Figure 2.
Sugarcane transport to trucks.

Sugarcane arrives at the sugar factory[18]

Sugarcane enters the mill[19]

Figure 3.
Trucks deliver sugarcane to factories.

expelled as bagasse waste [19]. Sugarcane harvesting age from planting to ready harvest is 12 months. Sugarcane is harvested by cutting sugarcane stalks at the bottom and top. Sugarcane of good quality for making sugar must be maintained during harvesting. The following figure shows the process that occurs in a sugar factory to produce sugar and remove the rest of the sugarcane process as bagasse [20–22].

After arriving at the factory, sugar cane is processed into white sugar or brown sugar with factory equipment as shown in **Figure 4** below. The end result of this sugarcane process is the result of sugarcane waste/bagasse [23, 24].

In its production process, sugarcane produces 90% bagasse, 5% molasses and 5% water [25]. The waste obtained from sugarcane during the process of producing sugar is known as bagasse which is used as fuel, material for paper pulp, organic fertilizer and animal feed. Not many industries have developed products made from bagasse. The panel board manufacturing industry and the bagasse fiber-reinforced asbestos-producing industry are the small industries that have started developing bagasse. Sugarcane is a plant used as a raw material in the production of sugar and Monosodium glutamate (MSG) [26–28].

A total of 32% bagasse is produced from the weight of milled sugarcane. It contains cellulose, lignin and hemicellulose compounds which are by-products of the sugarcane extraction process [29, 30]. As seen in the following **Figure 5**, from the observation of plant cell wall macrofibrils, there are three main components, namely lignin, cellulose and hemicellulose. In hemicellulose itself there are pentose and hexose [31–33].

Lignin, which mainly accumulates in plant stems, fills the space in the cell wall between cellulose, hemicellulose and pectin [34, 35]. This substance is present in all vascular plants but not in Bryophytes, which supports the idea that lignin's original function is limited to water transport. Furthermore, as one of the main components in bagasse, lignin is a complex polymer with high molecular weight composed of phenylpropane units which are the main component of wood building blocks. Lignin content is more in softwood compared to hardwood [36–38].

With a ligno-cellulose content, a fiber length of 1.7 to 2 mm and a diameter of about 20 micros, bagasse may actually be used as a raw material in chemical, petroleum, paper, brake canvas and mushroom industries. Therefore, it is economically utilized not only as a source of fuel energy in steam boilers, but also as a raw material for papermaking or a source of animal feed. In general, in Indonesia, sugar factories use bagasse as fuel after it undergoes a drying process. Another consideration used in selecting bagasse is because the sugarcane land is quite large, which is spread from Western to Eastern Indonesia, from North Sumatra, Palembang, Lampung, Java and Sulawesi, hence natural resources are readily available. This is

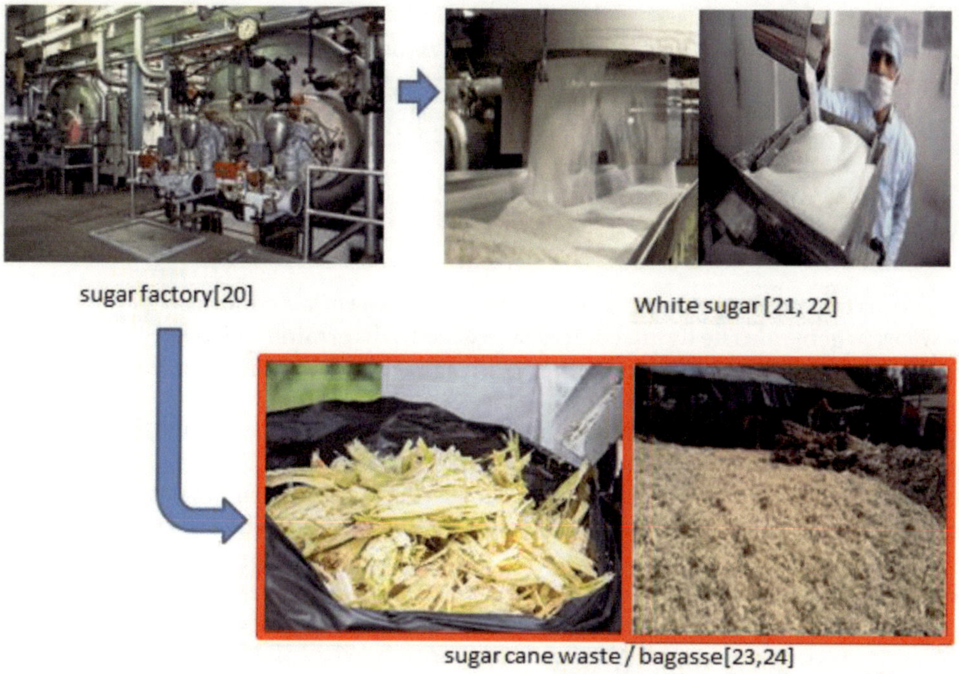

Figure 4.
The production process in a sugar factory.

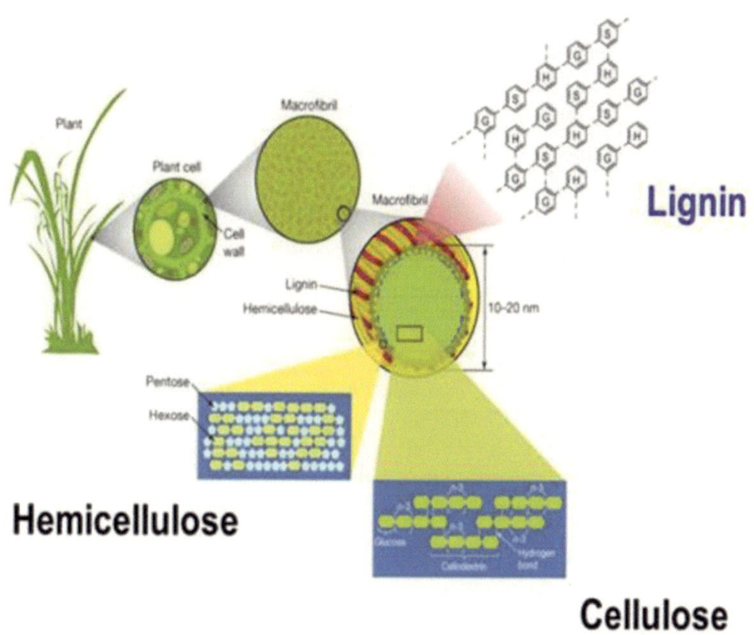

Figure 5.
Structure component of plant biomass [31].

also complemented by a plan of the local government to develop a sugar factory and sugarcane plantation. The development of sugarcane plantations supports the needs of the sugar industry, which in the making process will consequently produce a lot of waste. Each year, the amount of bagasse produced is quite abundant, easy to obtain, and cheap. Based on data from the Indonesian Sugar Plantation Research Center (P3GI), bagasse is obtained by 32% of milled sugarcane weight or around 10.2 million tons/year or per milled season throughout Indonesia [39, 40]. About 50% of the bagasse produced in each sugar factory is used as boiler fuel and the rest is dumped as waste which has low economic value [41, 42].

2. Methodology

The study method starts from a review of the potentials contained in various sources in general. The lignin content was used to determine the use of bagasse as a surfactant raw material. Several other wastes that also contain lignin, cellulose and hemicellulose are seen in **Table 1** below.

Sources of lignin in **Table 1** are broad petioles, needle stalks, leaves, corn cobs, peanut shells, wheat straw, bagasse, oil palm empty bunches. From the source of lignin, the focus of research is bagasse, because bagasse is sugarcane waste. The focus of this research is only on the bagasse. In **Table 1**, it can be seen that bagasse is a type of waste containing 25% lignin, 25% hemicellulose and 50% cellulose. By processing the bagasse, it can be used as a solution for sugarcane solid waste. Another consideration for the decision to use bagasse is the chemical composition of bagasse. The chemical composition of bagasse consists of ash, lignin, cellulose, extract, pentose and SiO_2. Its chemical composition is shown in **Table 2** below.

Bagasse is a vegetable source that contains large amounts of lignin which is a byproduct of the sugarcane liquid extraction process. This extraction produces bagasse about 32% of milled sugarcane weight Based on chemical analysis, the

No.	Waste	Cellulose (%)	Hemicellulose (%)	Lignin (%)
1.	Broad leaf trunk[41]	40–55	24–40	18–25
2.	Needle leaf trunk[41]	45–50	25–35	25–35
3.	Leaf[42]	15–20	80–85	0
4.	Corn cobs[42]	45	35	15
5.	Peanut shell[42]	23–30	25–30	30–40
6.	Wheat straw[43]	30	50	15
7.	Bagasse[43]	50	25	25
8.	Palm empty bunches[43]	41,30 – 46,50	25,30 – 33,80	27,50–32

Table 1.
Components of lignin in various vegetable wastes.

Content	Level (%)
Ash	3.82
Lignin	22.09
Cellulose	37.65
Extract	1.81
Pentosan	27.97
SiO_2	3.01

Table 2.
Chemical composition of bagasse [43].

average bagasse has a chemical composition of 3.28% ash, 22.09% lignin, 37.65% cellulose, 1.81% extract, 27.97% pentosan and 3.01% SiO2. The lignin content of 22.09% is a good potential for bagasse to be processed into lignosulfonates. Most of the content in bagasse is ligno-cellulose which is a natural polymer with high molecular weight that is rich in energy. Therefore, a large amount has the potential of been used as an energy source [43, 44].

One of the determining factors for the success of a livestock business is feed, where more than half of the production costs are used to fulfill feed needs. Therefore, efforts need to be made to ensure that provision of feed is cheap, easy to obtain and not competitive with human needs. Forage is one of the main foods for livestock, but its continuous supply has encountered several obstacles due to the narrow land available for forage cultivation which reduces the availability of feed. One alternative to overcome the problem of feed availability is to utilize agricultural byproducts [45]. According to Sajjad Karimi [46], the by-product of sugarcane milling may be used as animal feed because it is tolerant of summer, resistant to pests and diseases and easily available in the dry season when forage is less available [47]. The utilization of sugarcane as a feed ingredient requires technology because it has high crude fiber and low crude protein content. As a feed, several ingredients need to be added to bagasse in order to complement the mineral requirements needed in the feed material [48–51].

Bagasse waste has the opportunity to be optimally utilized as alternative energy which is beneficial to community needs and is friendly to the environment [52]. Biomass is a material obtained from plants directly or indirectly and is used as energy or materials in large quantities. Furthermore, it is known as "Fitomass" and is often translated as bioresource or resources obtained from living organisms. Biomass may actually be used directly without going through charcoal production first [53]. Sugarcane bagasse is generally used as boiler fuel to produce the energy needed in the sugar-making process which simultaneously produces a large amount of waste. Bagasse fiber is made up of cellulose which contains active carboxyl groups and lignin which contains phenolic groups. Several studies have utilized bagasse as an adsorbent for the removal of Congo Red dye [54] and reduction of iron content in well water [55]. Carbonized bagasse charcoal at 250°C for 2.5 hours [56] is used in the removal of heavy metals such as Pb, Cu, Cr and Cd. Furthermore, the effectiveness of this charcoal absorption has been tested [57] for the remediation of magnesium, manganese, zinc and nitrates in leachate. Based on the analysis result, the effectiveness of this charcoal is higher compared to the use of bagasse fiber. The use of this charcoal to change the characteristics of peat water and improve its quality into clean water has not been reported [58–60]. The adsorption process is one of the efforts made to improve the quality of pray water using bagasse charcoal as an adsorbent. The ability of bagasse biomass is increased by means of carbonization activation [61]. Furthermore, as an energy substitute, bagasse has been studied as bioethanol. From lignocellulosic biomass, ethanol can be produced [62–65]. For example, from bagasse hemicellulocicidrolysis, from the study of Canilha et al. reported the following results of 7.5 g / L, 0.30 g /g and 0.16 g/L [66]. The subject of discussion focuses on the characteristics and potential of lignocellulosic biomass, biomass conversion technology to ethanol and its potential development.

Study and development of science and technology in the field of making various composite materials to fulfill various purposes/needs have been widely carried out by educational and industrial circles. This study is reasonable due to the abundant availability of reinforcing fiber raw materials from organic composite reinforcing fibers such as bamboo, pineapple, sugarcane, banana and palm fibers [67] or inorganic reinforcing fibers and the quite high need/demand for processed composite materials on the market. Baggase fiber is one of the many natural fibers found in Indonesia. Post-harvest activities and processing of agricultural/plantation products, including the use of by-products and their processing residues, are still not optimal. In the sugarcane processing industry, the amount of bagasse produced may be up to 32% of each processed sugarcane. Till date, its use as raw material for the manufacture of particle boards, boiler fuel, organic fertilizers and animal feed is limited and has low economic value. Furthermore, the use of bagasse fiber as a reinforcing fiber for composite materials have very important meaning in terms of the utilization of industrial waste, especially the sugar-making industry in Indonesia which has not been optimized from an economic perspective and the utilization of its processed products. The results of this study are expected to bring new innovations in the development of non synthetic fiber-reinforced composite material technology. Therefore, bagasse fiber may be used as an alternative raw material, because it is easily obtained, almost available in all regions of Indonesia, a plantation crop widely cultivated by many farmers in Indonesia, more environmental friendly, a natural fiber and easier to process.

Bagasse is organic waste which is produced in many sugarcane processing factories in Indonesia. Bagasse is easy to obtain, does not endanger health and can be decomposed naturally (biodegradability). Therefore, the use of composite reinforcing fibers will be able to overcome environmental problems [68].

Several studies have also demonstrated the feasibility of using pre-treated bagasse at high consistency conditions to produce ethanol with a theoretical yield approaching 65% [69]. The production of bioethanol from bagasse can be obtained by engineering the evolution of the super active fermentation yeast *Saccharomyces cerevisiae* through the suppression gene deletion process. This may have detrimental effects on fermentation and the overexpression of all metabolic pathways for the fermentation of glucose, xylose, arabinose sugars and processes for higher tolerance to inhibitors [70]. There are many benefits that may be obtained from bagasse waste as shown in the following table (**Table 3**).

More study efforts are needed in relation to the use of bagasse as another product with more value, as shown in the following figure. In this **Figure 6**, it can be seen that the use of bagasse is applied to four major parts, namely for energy, biochemicals, food and materials. As an energy use, bagasse is used as charcoal, pellets, biogas, bioethanol, heat and electricity. As a biochemical, bagasse can be made for biopylomers, vanillin, enzymes, xylitol and furfural. Meanwhile, for food needs, bagasse can be made as protein, animal feed, fertilizer, soil conditioner. And finally, for material needs, bagasse can be used as an adsorbent, construction, textile fiber, boards and paper.

No.	Application	Reference
1.	Bagasse charcoal	[57, 61]
2.	Adsorbent	[55, 56]
3.	Organic composite reinforcement	[68, 71]
4.	Biomass	[53, 67]
5.	Animal feed	[46–52]
6.	Bioethanol	[69, 70]
7.	SLS surfactant	[72, 73]

Table 3.
Results of the study carried out on the benefits of this waste.

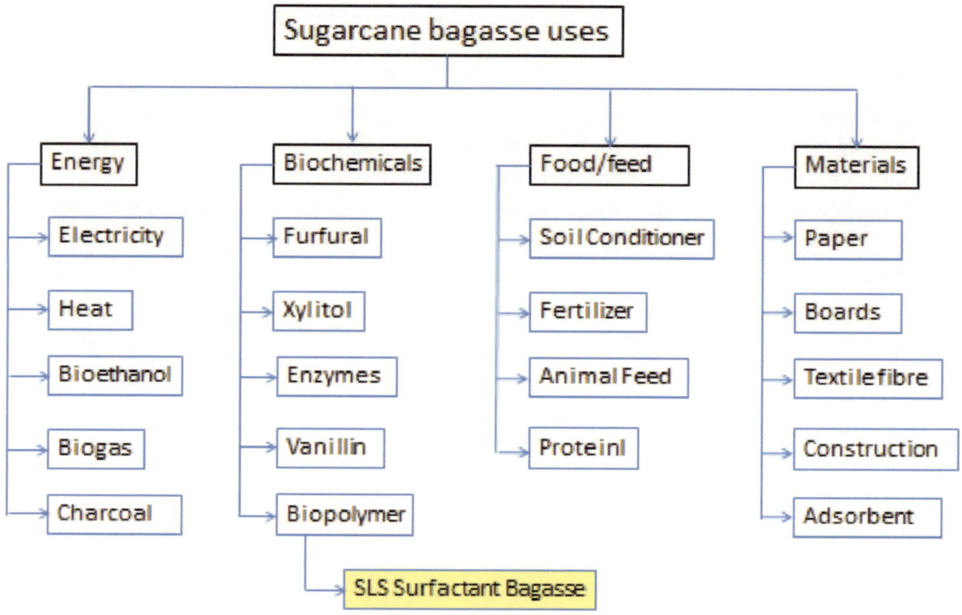

Figure 6.
Bagasse utilization diagram [74].

Several previous studies also stated that bagasse is a fiber containing cellulose with an active carboxyl group and lignin with a phenolic group. This research was conducted by Collepardi [72] and Setiati [73], processing bagasse into surfactant sodium lignosulfonate. This research is the manufacture of biopolymers which is part of biochemistry due to the large amount of lignin contained in bagasse. According to previous studies, the lignin content in bagasse was 25% as shown in **Table 1** above [75]. This type of surfactant is used as a raw material for fluid injection in oil fields to increase petroleum production. **Figure 6** above shows the position of the SLS surfactant bagasse synthesis which is part of the biopolymer.

The hoarding of bagasse as waste for a certain time cause problems, because this material is flammable, pollutes the surrounding environment and takes up a large area for storage [76]. So that this waste must be handled properly, by using it into other products.

The four main categories of bagasse application are currently being studied and utilized by the community. Various exploitation efforts continue to be made to minimize bagasse because sugarcane plantations and sugar factories are still developing [77]. Therefore, technological developments are still needed in processing and utilizing bagasse.

3. Results and discussion

One of the uses of bagasse is as a biopolymer, which is processed into Sodium LignoSulfonate (SLS) surfactant. SLS surfactant is an anionic surfactant, a type of surfactant that is widely used in the surfactant injection process in oil reservoirs. The function of this surfactant is to form a middle phase emulsion so that there is a decrease in interfacial tension (IFT) between the oil in the reservoir and the oil that sticks to the rock because it is difficult to move [78, 79]. With the availability of surfactants, the interfacial tension is low so that oil can be easily produced. The remaining oil in the reservoir can be removed and produced to increase oil recovery.

Processing of bagasse into SLS surfactant is carried out through two processes, namely hydrolysis and sulfonation [80–82]. Hydrolysis is the separation of lignin from bagasse by using a sieve analysis to obtain a size of 80 mesh. The lignin obtained is reacted with a sulfonation process using sodium bisulfite reagent to obtain sodium lignosulfonate surfactant [83]. The following figure shows the complete process from bagasse synthesis to SLS surfactant.

The synthesis process shown in **Figure 7** begins with the process of isolating lignin from bagasse. This process uses NaOH as a reagent, by heating for 5 hours at a temperature of 100°C. Then titrated using H_2SO_4 until a precipitate appears then filtered and dried, to become lignin. The lignin formed is then processed into lignosulfonate surfactants by a sulfonation process using $NaHSO_3$ as a reagent, reflux for 5 hours at a temperature of 150°C. As a result, it is dried to become lignosulfonate powder.

Lignosulfonates are lignin derivatives containing sulfonates that have hydrophilic groups which include sulfonate groups, phenyl hydroxyl and hydroxyl alcohol and hydrophobic groups (carbon chains) [84]. Therefore they are included in the anionic surfactant group.

The SLS surfactant produced from the synthesis of bagasse was tested for the components which it contains, using the FTIR (Fourier Transform Infra-Red) test. Tabulation of FTIR test results is seen in the following **Table 4**. Graph of FTIR test results can be seen in **Figure 8**.

From the **Figure 8** below, it is observed that the four clearest peaks show the absorption peak at wave number 1635.34 cm-1 as the stretching vibration region of alkene functional group –C=C– aromatic, wave number 1384.64 cm-1 as the stretching vibration region of sulfonate functional group S=O, wave number 1114.65 cm-1 as the bending vibration region of carboxylate functional group C=O and wave number 462,832 cm-1 as the bending vibration region of ester functional group S-OR.

The results of FTIR test showed that the lignosulfonate formed has components of alkenes, sulfates, carboxylic acids and esters. The synthesized SLS surfactant

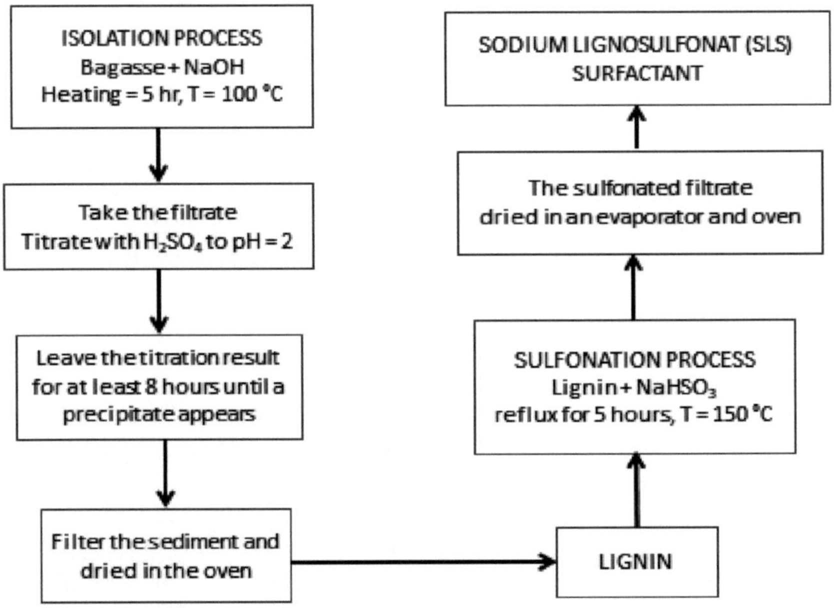

Figure 7.
Synthesis process of SLS surfactant from bagasse [73].

DOI: http://dx.doi.org/10.5772/intechopen.96373

No.	Component	Wave number(cm^{-1})
1.	Alkene C=C	1635.34
2.	Sulfonate S=O	1384.64
3.	Carboxylic Acids C=O	1114.65
4.	Ester S–OR	462.832

Table 4.
Spectrum of FTIR results of lignosulfonate from bagasse.

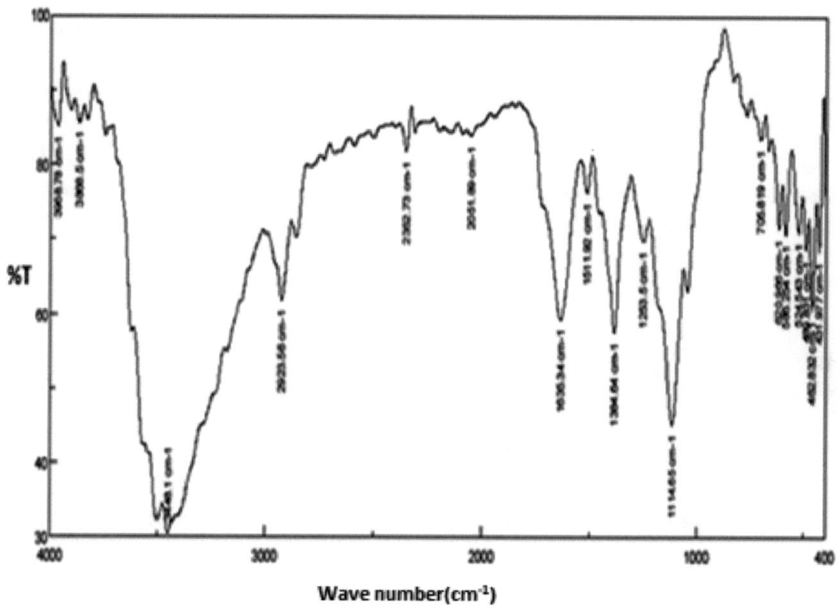

Figure 8.
FTIR test curve of SLS surfactant bagasse.

was compared to a commercial lignosulfonate surfactant, namely Patricia lignosulfonate. This surfactant consists of an alkene component with a wave number of 1630–1680 cm^{-1}, a sulfonate group with wave number of 1350 cm^{-1}, Carboxylic acids with wave number of 1000–1300 cm-1 and an ester with wave number of 500–540 cm^{-1} [85]. Compared to commercial lignosulfonate, SLS surfactant bagasse has a constituent component which is exactly the same as the comparator commercial lignosulfonate. There is only a slight difference in the absorption peak wave number detected, for example the alkene on SLS surfactant bagasse is 1635.34 cm^{-1}, which is still in the range of comparator lignosulfonate wave numbers of namely 1630–1680 cm^{-1}. Meanwhile, the sulfonate element in SLS surfactant bagasse is at a wave number of 1384.64 cm^{-1}, shifted slightly to the left compared to commercial lignosulfonate which has a sulfonate wave number of 1350 cm^{-1}, which indicates that this deviation only occurs by 2.56%. Furthermore, the carboxylic acids component with a wave number of 1114.65 cm^{-1} is still in the comparator lignosulfonate range, namely 1000–1300 cm^{-1}, while the last element (ester) provides a measurement result that deviates slightly to the right. Based on the results from the measurement of SLS surfactant bagasse, the ester is at a wave number of 462,832cm^{1}, while the comparator lignosulfonate has a wave number between 500 and 540 cm^{-1}. The deviation that occurs in this element is 7.4%.

The other indicator for the success of SLS surfactant bagasse as a raw material for surfactant injection in the EOR process may be seen from the HLB (Hydrophilic–Lipophilic Balance) value. HLB determination was used to determine the classification of the surfactant. The Myers table was the standard HBL value used in which the emulsion-forming fluid has an HLB value of 8–11 [86]. The components that needed to be known were the lipophilic and hydrophilic groups. This component is known from the atomic element which is measured based on the NMR (Nuclear Magnetic Resonance) test.

Based on the NMR spectrum analysis of the SLS surfactant bagasse sample, it turns out that the atomic number of C = 11, O = 8, H = 16 and S = 1. The molecular mass of the lignosulfonate monomer may be determined by looking at the presence of C, O, H and S atoms in their structure. Therefore, the empirical formula of the lignosulfonate monomer is $(C_{11}H_{16}O_8S)_n$, with a relative molecular mass of 308.06.

Furthermore, based on the NMR analysis, it is easy to identify which groups are classified as hydrophilic or lipophilic. The Lipophilic group consists of the elements ($= CH -$, $- CH_2 -$, $- CH_3$), while the hydrophilic group consists of the elements($= - SO_3Na$) and (-OH) elements. The grouping is seen in the **Table 5** below.

Based on the number of lipophilic and hydrophilic element atoms, this HLB value may be calculated using the equation as follows:

$$HLB = 20 * (M_h) / (M_l + M_h) \tag{1}$$

Where: M_h = the molecular weight of hydrophilic group.
M_l = the molecular weight of the lipophilic group

$$M_h = (SO_3Na) + (OH)x3 = (32 + 48 + 23) + 51 = 154 \tag{2}$$

$$M_l = (CH)x3 + (CH2)x3 + (CH3)x2 = 111 \tag{3}$$

$$HLB = 20 * (M_h) / (M_l + M_h) = 20 * 154 / (111 + 154)) = 11.62 \tag{4}$$

The result calculation, SLS surfactant bagasse has an HLB value of 11.62. Therefore, based on **Table 6**, it is suitable for use in the O/W (oil in water) emulsion type system, which means that the surfactant is soluble in water [86].

Furthermore, the HLB value was determined empirically with a scale of 0–20, as shown in **Table 6** below. The higher the HLB value, the more hydrophilic the surfactant and the more soluble it is in water, or known as an O/W (Oil in Water) emulsion. Meanwhile, lower HLB value indicates that the surfactant is a W/O

Classification	Group	Number
Lipophilic Groups	$=CH-$	3
	$-CH_2-$	3
	$-CH_3$	2
Hydrophilic Groups	$-SO_3Na$	1
	$-OH$	3

Table 5.
Grouping of bagasse NaLS surfactant functional groups.

HLB Value Range	Application
2–6	W/O emulsion
7–9	Wetting agent
8–18	O/W emulsion
3–15	Detergent
15–18	Solubilization

Table 6.
HLB value and its application [86].

(Water in Oil) emulsion, which will be more soluble in oil [86, 87]. Surfactants are usually amphiphilic organic compounds, which contain hydrophobic (tail) and hydrophilic (head). In addition, they spread in water and absorb at the interface between air and water or at the interface between oil and water.

Therefore, this SLS surfactant bagasse may be used as an injection fluid because it forms an O/W emulsion and is dissolved in water. In accordance with the injection mechanism, the surfactant is dissolved in the formation water and is injected into the reservoir to push the oil trapped in the rock pores. So, SLS surfactant bagasse has the potential of been used as a surfactant injection in the enhanced oil recovery process.

4. Conclusions

Based on studies conducted on bagasse and its application as a new, more useful product, it appears that the lignin content in bagasse has the potential to be processed into sodium lignosulfonate (SLS) surfactant. This SLS surfactant functions as an injection liquid to increase oil recovery. The laboratory test results corroborate these claims, which show that bagasse SLS surfactant as lignosulfonate has four main components, namely alkenes, sulfonic acids, carboxylic acids and esters. Furthermore, it has an HLB ((Hydrophilic–Lipophilic Balance) value of 11.62 which indicates that it functions as an emulsifier that dissolves in water and forms a surfactant solution. SLS surfactant bagasse forms an emulsion and reduces interfacial tension (IFT), so that oil granules are easier to produce. So it can be concluded that bagasse has good potential to be used as raw material for lignosulfonate surfactants.

Author details

Rini Setiati[1*], Aqlyna Fatahanissa[1], Shabrina Sri Riswati[1], Septoratno Siregar[2] and Deana Wahyuningrum[3]

1 Petroleum Engineering, Faculty of Earth Technology and Energy, Trisakti University, Jakarta, Indonesia

2 Petroleum Engineering, Faculty of Mining and Petroleum Engineering, Bandung Institute of Technology, Bandung, Indonesia

3 Chemistry, Faculty of Math and Science, Bandung Institute of Technology, Bandung, Indonesia

*Address all correspondence to: rinisetiati@trisakti.ac.id

IntechOpen

© 2021 The Author(s). Licensee IntechOpen. This chapter is distributed under the terms of the Creative Commons Attribution License (http://creativecommons.org/licenses/by/3.0), which permits unrestricted use, distribution, and reproduction in any medium, provided the original work is properly cited.

References

[1] Tajuddin Bantacut, 2010, Sugar Self-Sufficiency: Prospects and Strategies for Achievement, PANGAN, Vol. 19 No. 3 September 2010: 245-256, http://www.jurnalpangan.com/index.php/pangan/article/view/142

[2] Nkosingiphile Samuel Zulu, Melusi Sibanda and Bokang Stephen Tlali, 2019, Factors Affecting Sugarcane Production by Small-Scale Growers in Ndwedwe Local Unicipality, South Africa, Agriculture 2019, 9, 170; doi:10.3390/agriculture9080170

[3] Andi Amran Sulaiman, Yiyi Sulaeman, Novia Mustikasari, Dedi Nursyamsi and Andi Muhammad Syakir, 2019, Increasing Sugar Production in Indonesia through Land Suitability Analysis and Sugar Mill Restructuring, Land 2019, 8, 61; doi:10.3390/land8040061

[4] Gan-Lin Chen, Feng-Jin Zheng, Bo Lin,Tian-Shun Wang & Yang-Rui Li, 2013, Preparation and Characteristics of Sugarcane Low Alcoholic Drink by Submerged Alcoholic Fermentation, Sugar Tech (2013) 15:412-416, DOI 10.1007/s12355-013-0248-3

[5] Anika Knüppel, Martin J. Shipley, Clare H. Llewellyn & Eric J. Brunner, 2017, Sugar intake from sweet food and beverages, common mental disorder and depression: prospective findings from the Whitehall II study, Scientific Reports | 7:6287| DOI:10.1038/s41598-017-05649-7

[6] Subdirectorate of Estate Crops Statistics, 2018, Indonesian Sugar Cane Statistics 2018, BPS-Statistics Indonesia, BPS Catalogue: 5504004, https://knoema.com/atlas/Indonesia/topics/Agriculture/Crops-Production-Yield/Sugar-cane-yield

[7] Chinnaraja Chinnadurai, 2017, Potential Health Benefits of Sugarcane, Research Gate, DOI: 10.1007/978-3-319-58946-6_1

[8] Garusti, Yoga A. Yogi, Nurindah, 2019, Analysis of Tanjung Sugar Quality of Three Sugarcane Varieties, Jurnal Littri 25(2), Desember 2019. Hlm. 91-99, DOI: http://dx.doi.org/10.21082/littri.v25n2.2019. 91-99

[9] Nazmi Mat Nawi, Troy Jensen, and Guangnan Chen, 2012, The Application of Spectroscopic Methods to Predict Sugarcane Quality Based on Stalk Cross-sectional Scanning, Journal American Society of Sugar Cane Technologists, Vol. 32, 2012, https://core.ac.uk/download/pdf/11048657.pdf

[10] H Nurdin, H Hasanuddin, D Darmawi, F Prasetya, 2018, Analysis of Calorific Value of Tibarau Cane Briquette, OP Conf. Series: Materials Science and Engineering 335 (2018) 012058 doi:10.1088/1757-899X/335/1/012058

[11] Irvan Eka Kurniawan, Purwono, 2018, Harvest, Load and Transport in PGMadukismo Area, Yogyakarta, Bul. Agrohorti 6(3): 354-361

[12] Ahmad Thoriq, Wahyu K. Sugandi, Rizky Mulya Sampurno, Reza Permana Aji, 2017, Study of Cycle Time on Loading and Transporting Process in PG. Jatituju, Majalengka, West Java, Jurnal Teknotan Vol. 11 No. 1, April 2017, DOI : 10.24198/jt.vol11n1.7

[13] PD Riajaya, 2019, Rainyseason period and climate classification in sugarcane plantation regions in Indonesia, IOP Conf. Series: Earth and Environmental Science 418 (2020) 012058, doi:10.1088/1755-1315/418/1/012058

[14] Rachana Devkota, Laxmi Prasad Pant, Hom Nath Gartaula, Kirit Patel,Devendra Gauchan, Helen

Hambly-Odame, Balaram Thapa and Manish N. Raizada , 2020, Responsible Agricultural Mechanization Innovationfor the Sustainable Development of Nepal's HillsideFarming System, Sustainability 2020,12, 374; doi:10.3390/su12010374

[15] S. Shankar, C. Maheswari, R. Gowtham, P. Kiruba, K. Mohansrinivas 2019, Design And Fabrication Of PortableSugarcane Harvesting Machine, International Journal Of Scientific & Technology Research Volume 8, Issue 12, December 2019, http://www.ijstr.org/final-print/dec2019/Design-And-Fabrication-Of-Portable-Sugarcane-Harvesting-Machine.pdf

[16] H. A. Abdel-Mawla, 2014, Optimal design of the tractor-front-mounted sugarcane grab loader, Agric Eng Int: CIGR Journal Vol. 16, No.2, http://www.cigrjournal.org

[17] G.A. KENT, 2010, Design, Manufacturing And Maintenance Of Sugar Mill Equipment: Issct Engineering Workshop 2008 Summary Report, Proc. Int. Soc. Sugar Cane Technol., Vol. 27, 2010, https://core.ac.uk/download/pdf/33495672.pdf

[18] A Kustiyo, Y Arkeman, 2018, Design for improvement of sugar factory performance basedon statistical thinking, IOP Conf. Series: Earth and Environmental Science 335 (2019) 012033, doi:10.1088/1755-1315/335/1/012033

[19] Gunawan, T Bantacut, M Romli and E Noor, 2017, Biomass by-product from crystal sugar production: A comparative study between Ngadirejo and Mauritius sugar mill, IOP Conf. Series: Earth and Environmental Science 141 (2018) 012009, doi :10.1088/1755-1315/141/1/012009

[20] Mohd Parvez, 2017, Steam Boiler, Al Falah University, https://www.researchgate.net/publication/320057473_Steam_Boiler

[21] Widodo, D R Kamardiani and D F Pratiwi, 2019, Partnership between sugar company and small farmer on sugar cane farming in dry climate area of Dompu District, West Nusa Tenggara Province, Indonesia, IOP Conf. Series: Earth and Environmental Science423 (2020) 012029, doi:10.1088/1755-1315/423/1/012029

[22] Ju Young Lee, Rosamond L Naylor, Anjuli Jain Figueroa and Steven M Gorelick, 2020, Water-food-energy challenges in India: political economy of the sugar industry, Environ. Res. Lett.15(2020) 084020, https://doi.org/10.1088/1748-9326/ab9925

[23] Joner Oliveira Alves, Jorge Alberto Soares Tenório, Chuanwei Zhuo, Yiannis Angelo Levendis, 2012, Characterization of Nanomaterials Produced from Sugarcane Bagasse, Journal of Materials Research and Technology, https://core.ac.uk/download/pdf/82724939.pdf

[24] Daissy Lorena Restrepo-Serna, Jimmy Anderson Martínez-Ruano andCarlos Ariel Cardona-Alzate, 2018, Energy Efficiency of Biorefinery Schemes Using Sugarcane Bagasse as Raw Material, Energies 2018,11, 3474; doi:10.3390/en11123474

[25] Gunawan, Tajuddin Bantacut, Muhammad Romli and Erliza Noor, 2019, Life Cycle Assessment of Cane-sugar in Indonesian Sugar Mill: Energy Use and GHG Emissions, IOP Conf. Series: Materials Science and Engineering 536 (2019) 012059, doi:10.1088/1757-899X/536/1/012059

[26] Tushar Kanti Bera, Sanjit Kumar Kar, Prem Kumar Yadav, Prithwiraj Mukherjee, Shankar Yadav, Bishal Joshi, 2017, Effects of monosodium glutamate on human health: A systematic review, World Journal of Pharmaceutical Sciences, 2017; 5(5): 139-144, https://www.researchgate.net/publication/320191030

[27] Kalapanda M Appaiah, 2010, Monosodium Glutamate in Foods and its Biological Effects, DOI: 10.1016/B978-0-12-374845-4.00013-8

[28] Campbell, A. (2014). Monosodium Glutamate (MSG). Encyclopedia of Toxicology, 391-392. doi:10.1016/b978-0-12-386454-3.00040-3

[29] Afriandi, Fajril Akbar, Idral Amri, 2015, The Study of Making Oxalic Acid with Variations of Stirring Speed and Stirring Time of Sugarcane Base Material, Jom FTEKNIK Volume 3 No. 1 Februari 2015

[30] Cesar Benigno Granda Cotlear, 2004, Sugarcane Juice Extraction And Preservation, And Long-Term Lime Pretreatment Of Bagasse, Dissertation, Chemical Engineering Texas A&M University

[31] Rubin, 2008, Genomics of Celullosic Biofuels, Nature 454, 841-848, DOI: 10.1038/nature07190

[32] Lv, S., Yu, Q., Zhuang, X., Yuan, Z., Wang, W., Wang, Tan X. (2013). The Influence of Hemicellulose and Lignin Removal on the Enzymatic Digestibility from Sugarcane Bagasse. BioEnergy Research, 6(4), 1128-1134. doi:10.1007/s12155-013-9297-4

[33] Michel Brienzo, Ana Flavia Azevedo Carvalho, Franciane Cristina de Figueiredo and Pedro de Oliva Neto, 2016, Sugarcane Bagasse Hemicellulose Properties, Extraction Technologies And Xylooligosaccharides Production, https://www.researchgate.net/publication/309414077_SUGARCANE_BAGASSE_HEMICELLULOSE_PROPERTIES_EXTRACTION_TECHNOLOGIES_AND_XYLOOLIGOSACCHARIDES_PRODUCTION

[34] Rezende, C., de Lima, M., Maziero, P., deAzevedo, E., Garcia, W., & Polikarpov, I. (2011). Chemical and morphological characterization of sugarcane bagasse submitted to a delignification process for enhanced enzymatic digestibility. Biotechnology for Biofuels, 4(1), 54. doi:10.1186/1754-6834-4-54

[35] Teboho C. Mokhena, Mokgaotsa J. Mochane, Tshwafo E. Motaung, Linda Z. Linganiso, Oriel M. Thekisoe, 2018, Sugarcane Bagasse and Cellulose Polymer Composites, Intechopen, http://dx.doi.org/10.5772/intechopen.71497

[36] Nowshin Farjana Silvy, Md. Shamim Reza, Md. Nazim Uddin, Meghla Akther, 2018, Comparison between Different Components of Some Available Hardwood and Softwood in Bangladesh, IOSR Journal of Biotechnology and Biochemistry (IOSR-JBB), Volume 4, Issue 1(Jan.-Feb2018), PP 01-05, http://www.iosrjournals.org/iosr-jbb/papers/Volume%204,%20Issue%201/A0401010105.pdf

[37] Novaes, E., Kirst, M., Chiang, V., Winter-Sederoff, H., & Sederoff, R. (2010). Lignin and Biomass: A Negative Correlation for Wood Formation and Lignin Content in Trees. PLANT PHYSIOLOGY, 154(2), 555-561. doi:10.1104/pp.110.161281

[38] Ricardo B. Santos, Ewellyn A. Capanema, Mikhail Yu. Balakshin, Hou-min Chang, Hasan Jameel, 2012, Lignin Structural Variation in Hardwood Species, Journal of Agricultural and Food Chemistry · April 2012, DOI: 10.1021/jf301276a

[39] Maryudi, Agus Aktawan, Siti Salamah, 2018, Conversion of Biomass of Bagasse to Syngas through Downdraft Gasification, Jurnal Bahan Alam Terbarukan, JBAT 7(1) (2018) 28-33, DOI 10.15294/jbat.v7i1.11621

[40] Lely Meilani , Kiagus. A Roni, Atikah, Netty Herawati, 2019, Conversion Of Sugar Cane Bagasse

become Bioethanol Using Variations Of Yeast Types And Fermentation Temperatures With Socarification Process, Journal of Computational and Theoretical Nanoscience · August 2019, http://DOI:10.1166/jctn.2019.8581

[41] Reshamwala, S., Shawky, B.T. and Dale, B.E. (1995). Ethanol production from enzymatic hydrolysates of AFEXtreated coastal Bermuda grass and switchgrass. Applied Biochemistry Biotechnology. 51/52, 43-55, https://agris.fao.org/agris-search/search.do?recordID=US9603400

[42] Pandey, Ashok., Carlos R. Soccol, Poonam Nigam, Vanete T. Soccol. (2000). Biotechnological potential of agro-industrial residues. I: sugarcane bagasse. Bioresource Technology. 74: 69-80, https://doi.org/10.1016/S0960-8524(99)00142-X

[43] Naomi P. Carnaje ID, Romel B.Talagon, JoseP. Peralta, KalpitShah, JorgePaz-Ferreiro ID, 2018, Development and characterization of charcoal briquettes from water hyacinth (Eichhorniacrassipes)-molasses blend, PLOS ONE November 9, 2018, https://doi.org/10.1371/journal.pone.0207135

[44] M Cholid and MMachfud , 2019, Acceleration adaption for integrated sugarcane cultivation technology through field laboratory practices, IOP Conf. Series: Earth and Environmental Science418 (2020) 012059, http://doi:10.1088/1755-1315/418/1/012059

[45] Samadi, Sitti Wajizah, Yunasri Usman, Denny Riayatsyah, Zidny Al-Firdausyi, 2016, Improving Sugarcane Bagasse as Animal Feed by Ammoniation and Followed by Fermentationwith Trichoderma harzianum(In Vitro Study), Animal Production, Vol 18, No 1 (2016), https://animalproduction.net/index.php/JAP/article/view/516/463

[46] Sajjad Karimi, Nasrollah Mahboobi Soofiani, Amir Mahboubi and Mohammad J. Taherzadeh, 2018, Use of Organic Wastes and Industrial By-Products to Produce Filamentous Fungi with Potential asAqua-Feed Ingredients, Sustainability 2018,10, 3296; http://doi:10.3390/su10093296

[47] Ratchataporn Lunsin, Suntariporn Duanyai, Ruangyote Pilajun, Somporn Duanyai, Prapatsorn Sombatsri. 2018, Effect of urea- and molasses-treated sugarcane bagasse on nutrientcomposition andin vitrorumen fermentation in dairy cows, Agriculture and Natural Resources 52 (2018) 622-627, http://doi.org/10.1016/j.anres.2018.11.010

[48] E.M. Shahowna1, A.G. Mahala, A.M. Mokhtar, E.O. Amasaib and Balgees Attaelmnan, 2013, Evaluation of nutritive value of sugar cane bagassefermented with poultry litter as animal feed, African Journal of Food Science and Technology (ISSN: 2141-5455) Vol. 4(5) pp. 106-109, May, 2013, http://www.interesjournals.org/AJFST

[49] Sotiris I. Patsios, Anna Dedousi, Evangelia N. Sossidou, Antonios Zdragas, 2020, Sustainable Animal Feed Protein through the Cultivation of YARROWIA Lipolyticaon Agro-Industrial Wastes and by-Products, Sustainability 2020,12, 1398; doi:10.3390/su12041398

[50] B. N. Suresh and B. S. V. Reddy, 2011, Dried Sugarcane Press Residue as a Potential Feed Ingredient Source of Nutrients for Poultry, Asian Australasian Journal of Animal Sciences· November 2011 Vol. 24, No. 11 : 1595-1600, DOI: 10.5713/ajas.2011.11054

[51] Anukam, A., Mamphweli, S., Reddy, P., Meyer, E., & Okoh, O. (2016). Pre-processing of sugarcane bagasse for gasification in a downdraft biomass gasifier system: A comprehensive review. Renewable and Sustainable Energy Reviews, 66, 775-801. doi:10.1016/j.rser.2016.08.046

[52] Chukwuneke J. L,*, Umeji A. C., Sinebe J. E., Fakiyesi O. B, 2020, Sugarcane Industry Waste Recovery: A Case StudyUsing Thermochemical Conversion Technologies toIncrease Sustainability, Appl. Sci. 2020,10, 6481; doi:10.3390/app10186481

[53] Mochamad Syamsiro, 2016, Improving Biomass Solid Fuel Quality With Densification and Torrefaction Processes , J. Mek. Sist. TermalVol. 1(1)2016:7-13, https://core.ac.uk/download/pdf/231290149.pdf

[54] Zhang, Z., Monghaddam, L., O'Hara,M.I., dan Doherty S..O.W. 2011. Congo Red Adsorption by ball milled sugarcane bagasse. Chemical Engineering Journal. 178:122-128. https://www.sciencedirect.com/science/article/abs/pii/S1385894711012630

[55] Thaisa Caroline Andrade Siqueira, Isabella Zanette da Silva, Andressa Jenifer Rubio, Rosângela Bergamasco, Francielli Gasparotto, Edneia Aparecida de Souza Paccola and Natália Ueda Yamaguchi, 2020, Sugarcane Bagasse as an Efficient Biosorbent forMethylene Blue Removal: Kinetics, Isothermsand Thermodynamics, Int. J. Environ. Res. Public Health 2020,17, 526, doi:10.3390/ijerph17020526

[56] Hala Ahmed Hegazi, 2013, Removal of heavy metals from wastewater usingagricultural and industrial wastes as adsorbents, Housing and Building National Research Center, HBRC Journal 9 : 276-282, http://dx.doi.org/10.1016/j.hbrcj.2013.08.004

[57] Justine Barthod & Cornelia Rumpel & Marie-France Dignac , 2018, Composting with additives to improve organic amendments. A review, Agronomy for Sustainable Development(2018) 38: 17, https://doi.org/10.1007/s13593-018-0491-9

[58] Silvio Rainho Teixeira, Agda Eunice de Souza, Angel Fidel Vilche Peña, Regiane Godoy de Lima and Álvaro Gil Miguel, 2011, Use of Charcoal and Partially Pirolysed Biomaterial in Fly Ash to Produce Briquettes: Sugarcane Bagasse, Intechopen Book Chapter, DOI: 10.5772/20505

[59] Baker DA, Rials TG (2013). Recent advances in low-cost carbon fiber manufacture from lignin. J Appl Polym Sci 130(2):713-728. https://doi.org/10.1002/app.39273

[60] Milbrandt A, Booth S, (2016), Carbon fiber from biomass. National Renewable Energy Laboratory (NREL), Golden, https://doi.org/10.2172/1326730

[61] Qing Xu, Tao Ji, San-Ji Gao, Zhengxian Yang and Nengsen Wu, 2018, Characteristics and Applications of Sugar CaneBagasse Ash Waste in Cementitious Materials, Materials 2019,12, 39; doi:10.3390/ma12010039

[62] Sandra Cerqueira Pereira, Larissa Maehara, Cristina Maria Monteiro Machado and Cristiane Sanchez Farinas, 2015, 2G ethanol from the whole sugar cane lignocellulosic biomass, Biotechnology for Biofuels (2015) 8:44, DOI 10.1186/s13068-015-0224-0

[63] Tyagi, S., Lee, K.-J., Mulla, S. I., Garg, N., & Chae, J.-C. (2019). Production of Bioethanol From Sugarcane Bagasse: Current Approaches and Perspectives. Applied Microbiology and Bioengineering, 21-42. doi:10.1016/b978-0-12-815407-6.00002-2.

[64] Cerqueira Leite RC, Leal MRLV, Cortez LAB, Griffin WM, Scandiffio MIG. Can Bra-zil replace 5% of the 2025 gasoline world demand with ethanol?. Energy 2009; 34655-661. https://doi.org/10.1016/j.energy.2008.11.001

[65] Bothast, R. J., & Saha, B. C. (1997). Ethanol Production from Agricultural Biomass Substrates. Advances in Applied Microbiology, 261-286. doi:10.1016/s0065-2164(08)70464-7

[66] Larissa Canilha, Rita de Cássia Lacerda Brambilla Rodrigues, Felipe Antônio Fernandes Antunes, Anuj Kumar Chandel, Thais Suzane dos Santos Milessi, Maria das Graças Almeida Felipe andSilvio Silvério da Silva, 2013, Bioconversion of Hemicellulose fromSugarcane Biomass Into Sustainable, IntechOpen, http://dx.doi.org/10.5772/53832

[67] Hartono Yudo, Sukanto Jatmiko, 2008, Technical Analysis of Mechanical Strength of Composite Materials Reinforced Sugarcane (Baggase) in terms of tensile and impact strength, KAPAL, Vol. 5, No.2, Juni 2008, DOI: https://doi.org/10.14710/kpl.v5i2.3197

[68] M. Jawaida, H.P.S. Abdul Khalilb, Azman Hassana, Rudi Dunganic, A. Hadiyane, 2012, Effect of jute fibre loading on tensile and dynamic mechanical propertiesof oil palm epoxy composites, Journal Composites: Part B , http://dx.doi.org/10.1016/j.compositesb.2012.04.068

[69] Isis Amores, Ignacio Ballesteros, Paloma Manzanares, Felicia Sáez, Georgina Michelena, Mercedes Ballesteros, 2017, Ethanol Production from Sugarcane Bagasse Pretreated by Steam Explosion, Electronic Journal of Energy & Environment, DOI: 10.7770/ejee-V1N1-art519

[70] Monirul Islam Miskat, Ashfaq Ahmed, Hemal Chowdhury, Tamal Chowdhury, Piyal Chowdhury, Sadiq M. Sait and Young-Kwon Park , 2020, Assessing the Theoretical Prospects of Bioethanol Production as a Biofuel from Agricultural Residues in Bangladesh: A Review, Sustainability 2020,12, 8583; doi:10.3390/su12208583

[71] Zhang, Z., Monghaddam, L., O'Hara,M.I., dan Doherty S..O.W. 2011. Congo Red Adsorption by ballmilled sugarcane bagasse. Chemical Engineering Journal. 178:122-1., https://doi.org/10.1016/j.cej.2011.10.024

[72] Collepardi, (2005) , Mechanisms Of Action Of Superplasticizers, Proceedings of Second International Symposium on Concrete Tecnology for Sustainable February - Development with Emphasis on Infrastructure, Hyderabad, India, 27 February– 3 March (2005) pp. 527-541, http://www.encosrl.it/OLDSITE/pubblicazioni-scientifiche/pdf/additivi/92.pdf

[73] R Setiati, S Siregar, T Marhaendrajana, D Wahyuningrum, A Listyani, 2019, Sustainable Green Chemical Processing of Surfactant Synthesized from Bagasse for Enhanced Oil Recovery using Microwave Radiation, IOP Conf. Series: Materials Science and Engineering 662 (2019) 062005, doi:10.1088/1757-899X/662/6/062005

[74] Elias Martinez-Hernandez, Myriam Adela Amezcua-Allieri, Jhuma Sadhukhan and Jorge Aburto Anell (December 20th 2017). Sugarcane Bagasse Valorization Strategies for Bioethanol and Energy Production, Sugarcane - Technology and Research, Alexandre Bosco de Oliveira, IntechOpen, DOI: 10.5772/intechopen.72237.

[75] Syafwina HY, Watanabe T, Kuwahara M, (2002), Pre-treatment of oil palm empty fruit bunch by white-rot fungi for enzymatic saccharification, Wood Res 89 : 19-20, http://hdl.handle.net/2433/53108

[76] Lavarack, B.P., Griffin, G.J., Rodman, D., (2002). The acid hydrolysis of Sugarcane Bagasse Hemicellulose to Produce Xylose, Arabinose, Glucose and Other Products. Biomass & Bioenergy, 23: 367-380. https://doi.org/10.1016/S0961-9534(02)00066-1

[77] A.S. Dwi Saptati Nur Hidayati, Silva Kurniawan, Nalita Widya Restu, Bambang Ismuyanto, 2016, The Potential of Sugarcane as an Alternative for Raw Materials for Making Activated

Carbon, NATURAL B, Vol. 3, No. 4, Oktober 2016, pp 311-317, https://natural-b.ub.ac.id/index.php/natural-b/article/view/337

[78] Sheng JJ, 2015, Status of surfactant EOR technology, Petroleum Volume 1, Issue 2, June 2015, Pages 97-105, https://doi.org/10.1016/j.petlm.2015.07.003

[79] Bera A, Mandal A, 2015, Microemulsions: a novel approach to enhanced oil recovery: a review, Journal of Petroleum Exploration and Production Technology volume 5, pages 255-268(2015), https://link.springer.com/article/10.1007/s13202-014-0139-5

[80] Rini Setiati, Septoratno Siregar and Deana Wahyuningrum (November 19th 2020). Laboratory Optimization Study of Sulfonation Reaction toward Lignin Isolated from Bagasse [Online First], IntechOpen, DOI: 10.5772/intechopen.93662. Available from: https://www.intechopen.com/online-first/laboratory-optimization-study-of-sulfonation-reaction-toward-lignin-isolated-from-bagasse

[81] Lisna Efiyanti, Dian Anggraini Indrawan, Zulhan Arif, Devandri Hutapea, Ane Dwi Septina, 2020, Synthesis and Application of a Sulfonated Carbon Catalyst for a Hydrolisis Reaction, Indonesian Journal of Science & Technology 5(3) (2020) 410-420, http://ejournal.upi.edu/index.php/ijost

[82] Rini Setiati, Septoratno Siregar, Taufan Marhaendrajana, Deana Wahyuningrum, Sugiatmo Kasmungin, 2020, Green Chemistry Processing Of Surfactant Synthesized From Bagasse Using Microwave Radiation, International Journal Of Scientific & Technology Research Volume 9, Issue 03, March 2020, http://www.ijstr.org/final-print/mar2020/Green-Chemistry-Processing-Of-Surfactant-Synthesized-From-Bagasse-Using-Microwave-Radiation.pdf

[83] Rini Setiati, Septoratno Siregar, Taufan Marhaendrajana and Deana Wahyuningrum (August 23rd 2019). Surfactant Flooding for EOR Using Sodium Lignosulfonate Synthesized from Bagasse, Enhanced Oil Recovery Processes - New Technologies, Ariffin Samsuri, IntechOpen, DOI: 10.5772/intechopen.88689. Available from: https://www.intechopen.com/books/enhanced-oil-recovery-processes-new-technologies/surfactant-flooding-for-eor-using-sodium-lignosulfonate-synthesized-from-bagasse

[84] Nguyen Truong Giang, Tran Trung Kien, Nguyen Thi Hoa, Pham Van Thiem, 2016, A New Synthesis Process Of Lignosulfonate Using Lignin Recovered From Black Liquor Of Pulp And Paper Mills, Journal of Science and Technology 54 (4B) (2016) 1-10, DOI: 10.15625/2525-2518/54/4B/12017

[85] Patricia, R. J. (2009). Relationship between the structure of Fe-Lignosulfonate complexes determined by FTIR spectroscopy and their reduction by the leaf Fe reductase. The Proceedings of the International Plant Nutrition Colloquium XVI (hal. 1). Davis: University of California. https://escholarship.org/uc/item/9k69q71d

[86] Myers, D. (2006). Surfactant Science and Technology. New Yersey: Wiley Interscience. http://press.crosa.com.tw/wp-content/uploads/2016/11/Surfactant-Science-and-Technology-Third-Edition.pdf

[87] Akzo N. HLB and Emulsi-ficationSurface Chemistry LLC. 2011. Available from: http://surface.akzonobel.com